国家林业和草原局普通高等教育"十三五"规划教材

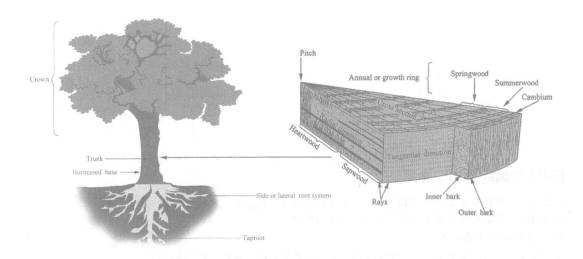

木材科学与工程专业英语

丁　涛 / 主　编

石江涛　詹天翼　张海洋 / 副主编

蔡力平 / 主　审

中国林业出版社

图书在版编目（CIP）数据

木材科学与工程专业英语 / 丁涛主编. —北京：中国林业出版社，2019.8（2025.3 重印）
国家林业和草原局普通高等教育"十三五"规划教材
ISBN 978-7-5219-0208-2

Ⅰ. ①木… Ⅱ. ①丁… Ⅲ. ①木材学—英语—高等学校—教材 Ⅳ. ①S781

中国版本图书馆 CIP 数据核字（2019）第 172622 号

国家林业和草原局生态文明教材及林业高校教材建设项目

中国林业出版社·教育分社

策划、责任编辑：杜　娟
电　　话：（010）83143553　　　传　　真：（010）83143516

出版发行：中国林业出版社（100009　北京西城区德内大街刘海胡同 7 号）
　　　　　E-mail: jiaocaipublic@163.com　　电话：（010）83143500
　　　　　http://www.forestry.gov.cn/lycb.html

经　销：	新华书店
印　刷：	北京中科印刷有限公司
版　次：	2019 年 8 月第 1 版
印　次：	2025 年 3 月第 2 次印刷
开　本：	850mm×1168mm　1/16
印　张：	5.75
字　数：	200 千字
定　价：	30.00 元

未经许可，不得以任何方式复制或抄袭本书之部分或全部内容。

版权所有　侵权必究

前　言

本教材的编写，是为了满足我国木材科学与产业在快速国际化进程中对英语语言技能的广泛需求。教材以"专业"和"英语"为两个支撑点，在内容编排上以木材科学与工程本科专业的主干课程结构为依据，涵盖全球木质资源概况、木材构造、主要物理性能、木材加工、结构与非结构人造板、木材改性与生物质能源等专业领域，具体选材则根据学科和产业发展的动态，注重内容的新颖性与时效性。在讲解上着眼于语言课程教学的重点，侧重英语词汇和用法，重点词汇的选择与解释主要参考美国材料试验协会（American Society for Testing and Materials）的相关术语标准（详见第 3 章），用法的说明则综合参考《牛津英语用法指南》（*Practical English Usage*）和《芝加哥格式手册》（*The Chicago Manual of Style*）等权威语言著作，以使读者通过学习掌握主要专业术语的英语表达和规范应用。

本书由南京林业大学丁涛（第 1、8 章）主编，南京林业大学石江涛（第 2、7 章）、詹天翼（第 3、4 章）和张海洋（第 5、6 章）任副主编。丁涛负责全文统稿，北德克萨斯州立大学蔡力平担任本书主审，书中原创插图由南京林业大学孙香绘制。本书由江苏省品牌专业（木材科学与工程）建设项目资助。我们谨向为本书写作、编辑、出版和发行等做出积极贡献的专家和出版工作者表示衷心的感谢！

本书主要面向木材科学与工程专业的本科教学，也可作为木材加工产业技术人员在资料翻译和对外贸易过程中的语言参考书。对于有兴趣深入学习的读者，本书在主要知识点说明中也给出了引申学习所需的参考文献。本书相关数字资源可扫描下方二维码。

兼顾业务与语言两方面的专业性是本书编写的一大挑战，编者在编写过程中进行了广泛的参考与细致审核，力求内容的科学性与表达的规范性，但不足之处再所难免，希望广大读者批评指正，以期将本书不完善。

本书数字资源

丁　涛
2019 年 5 月

目 录

Forest Resources and Wood Utilization 1

Wood Structure and Chemistry 10

Wood-Moisture Relations and Properties of Wood 27

Wood Products Manufacturing Process 40

Structural Panels 49

Nonstructural Panels 60

Wood Modification 69

Wood for Energy Production 79

Forest Resources and Wood Utilization

Global forest resources

Forests play a critical role in guaranteeing the well-being of humankind as they provide food, energy, shelter, and **fiber**, generate income and employment to allow communities and societies to prosper, and harbor biodiversity. They support sustainable agriculture by stabilizing soils and climate and regulating water flows. Last but not least, forests supply **wood**, which is one of the most important engineering materials for us.

The human understanding of global forest resources didn't exist until 1923 when the first comprehensive global forest resources assessment was conducted. The assessment, entitled *Forest Resources of the World*, did not actually include all countries and only around 3 billion **hectares** of forest area were reported. According to the *Global Forest Resources Assessment 2015*, the world has 4-billion-hectare forest, covering 30.6% of the total land area (Fig.1.1). Natural forest area accounts for 93% of global forest area while the rest belongs to planted forests or **plantations**. From 1990 to 2015, 93 countries in the world recorded net forest losses (totaling 242 million hectares), while 88 countries had net gains in forest area (totaling almost 113 million hectares). The world forest suffered a net loss of some 129 million ha. That means 1% fall of the global forest cover and is about the size of South Africa. Large-scale commercial agriculture is the main human practice responsible for forest loss, causing about 40% of **deforestation** in the tropics and subtropics. Planted forest area, on the other hand, has increased by over 113 million ha since 1990 thanks to the **afforestation** policies in many countries. The regional distribution of the net increase in forest area varies across regions. In Asia, 24 countries experienced a net increase of 73.1 million hectares in the forest area in the period 1990—2015, which was mainly due to large-scale afforestation programs in China.

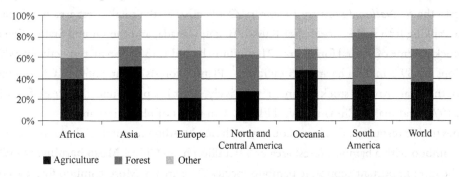

Fig.1.1 Land area by major land-use class (source: FAO, 2016)

As the world's forests continue to decline, more and more attention has been paid to sustainable forest management (SFM). More lands are designated as permanent forest, more assessments, monitoring, reporting, planning, and stakeholder involvement are taking place, and the legal frameworks for SFM are being widely adopted. Larger areas are being designated for the conservation of biodiversity while at the same time forests are meeting increasing demands for forest products and services.

Independent certification of forest management was first introduced in the late 1990s as a voluntary tool to promote SFM and trade of products coming from sustainably managed forests. Two international certification schemes have been widely accepted: the **Forest Stewardship Council** (FSC) certification and the **Programme for the Endorsement of Forest Certification** (PEFC) (Fig.1.2). Both schemes include criteria for best practices in forest management, covering environmental, social and economic aspects. The area covered by these two forest management certification schemes increased from 14 million ha in 2000 to over 438 million ha in 2014, of which, 58% is under PEFC and 42% is under FSC. These figures contain some double accounting (approximately 2%) as there are some forest management units that are certified by both schemes.

Fig.1.2 Logos of Forest Stewardship Council (left) and Programme for the Endorsement of Forest Certification (right)

China's forest resources

China is a country with rich forest resources. Forests once covered as much as 60% of the land area of the country. But that figure fell to as low as less than 10% by 1949 due to continuous forest conversion. Many efforts have been made ever since to recover the forests. The forest cover is now estimated at 21.6% (Fig.1.3), and the total **growing stock** volume at 16.4 billion m^3. Forest area, forest cover and growing stock volume have increased significantly thanks to continuous afforestation and natural forest protection. According to the *8th National Forestry Inventory (2009—2013)*, the total forest area in China is 207.7 million ha, ranking fifth in the world after Russia, Brazil, Canada, and the United States. The area and growing stock volume of natural forests are 121.8 million ha and 12.3 billion m^3 respectively, while those of plantations are 69.3 million ha and 2.5 billion m^3. Planted forests occupy 36% of China's forest area and 17% of the total growing stock volume respectively. That is a pronounced increase from 20% and 2% in the 1970s and makes China the world's No. 1 in terms of plantation area.

Besides timber resources, China is a country with rich **bamboo** resources. There are more than 500 species of bamboo with a bamboo forest area of 6.0 million ha in China. **Moso bamboo** (*Phyllostachys edulis*) is the most important economic bamboo species. The area of Moso bamboo forest accounts for 74% of the total bamboo forest area.

China's forest is primarily distributed in the state-owned forest area in nine provinces and collectively owned forest area in ten provinces. Timber forests are also primarily distributed in these regions (Fig.1.3). They collectively provide 87.7% of forest area and 83.9% of forest stock in China.

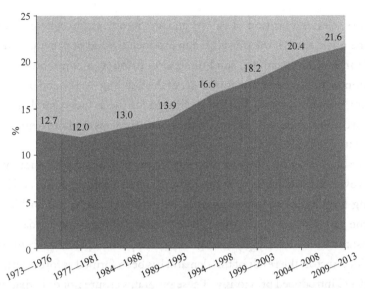

Fig.1.3 Change of China's forest coverage (source: State Forestry Administration, 2014)

Although China has made huge progress in forest recovery, it is still a country suffering shortages of forest resources. The forest area per capita is only about one-fourth of the world average. Forest stock per capita is only one-seventh of the world average. Over 50% of the market demand for wood relies on imports from abroad.

China's timber production increased significantly over the period from 1975 to 1995 and peaked at 677 million m^3 in 1995. Production has been declining since then to respond to the protection policies of natural forests. Nowadays, the average annual domestic timber supply is around 334 million m^3, with more and more timber produced from planted forests.

Wood as a sustainable biomaterial

Wood is one of the most valuable products offered by forests. Wood-based products are of great importance to our economy and living standard. It has long been the main raw material used to build houses, bridges, and boats. It is still among the first choices to make musical instruments, furniture, and flooring. And it will most probably be selected to produce alternatives to the fossil fuel-based products. Wood has many favorable characteristics, including low embodied energy, low carbon impact, and sustainability.

Embodied energy refers to the quantity of energy required to harvest, mine, manufacture, and transport materials or products to the point of use. Wood has a low level of embodied energy relative to many other materials used in construction, such as steel, concrete, aluminum, or plastic. Take the U. S. for example, over half the energy consumed in manufacturing wood products is from biomass and is typically produced from tree bark, **sawdust**, and by-products of pulping in the papermaking process. The U. S. wood product industry is the nation's leading producer and consumer of bioenergy, accounting for about 60% of its energy needs. Solid-sawn wood products have the lowest level of embodied energy; wood products requiring more processing steps require more energy to produce but still require significantly less energy than their non-wood counterparts.

Forests are the main carbon sink on earth as trees absorb CO_2 for photosynthesis. A tree that remains in the forest and dies releases a portion of its carbon back into the atmosphere as the woody material decomposes. On the other hand, if the tree is used to produce a wood or paper products, these products store carbon while in use. For example, solid dimension **lumber**, a common wood product used in building construction, sequestered carbon for the life of the building. At the end of a building's life, wood can be recovered for re-use in another structure, chipped for use as fuel or mulch, or sent to a landfill. If burned, the stored carbon is released, essentially the reverse process of photosynthesis. That is generally held as a carbon neutral process.

Wood is renewable, i.e., a flow of wood products can be maintained continuously with proper forest management. However, the sustainability of this resource requires afforestation and harvesting practices that ensure the long-term health and diversity of forests. Unfortunately, sustainable practices were not always applied in the past. Nor are they universally applied around the world today. Architects, product designers, and consumers are increasingly asking for certified wood products from sustainable sources. For the forest products sector, the result of this demand has been formed as forest certification programs, such as FSC and PEFC introduced previously. These programs ensure not only that the forest resources are harvested in a sustainable fashion but also that issues of biodiversity, habitat protection, and indigenous peoples' rights are included in land management plans.

Gradual reduction of **old-growth** or primary forests around the world has reduced the supply of large **logs** for lumber and veneer manufacturing. **Second-growth** wood from planted forests has emerged as an important alternative to fill the ever-increasing needs for wood. After proper modification or reconstruction, the performance of second-growth wood is as good as, and, in many cases, superior to the traditional one. In 2015, global industrial roundwood production was 1.8 billion m³ (Fig.1.4). At the country level, the five largest producers of roundwood are the U.S., Russia, China, Canada, and Brazil. The combined roundwood output of these countries was 1.0 billion m³ in 2015, or 55% of the world total. In 2015, global sawn-wood production was 452 million m³ (Fig.1.4), and the five largest producers are the U. S. , China, Canada, Russia, and Germany.

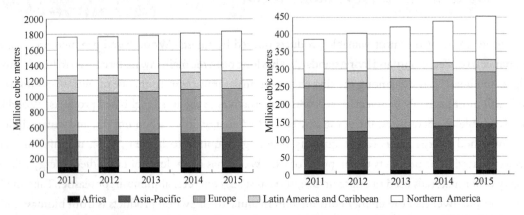

Fig.1.4　Global roundwood production (left) and sawn-wood production (right) (source: FAO, 2016)

Wood utilization and technology

Since the early days of human history, wood has been used for heating, building houses and extracting chemicals and other applications. But the large-scale use of wood as an engineering material didn't begin until about 150 years ago, when steam power and large lumber-manufacturing equipment were available.

The first type of saw to which mechanical power was applied was the old-fashioned "gate saw", also known as the frame or sash saw. To increase the efficiency of these saws, they were arranged in gangs to make a number of cuts at one pass of the log. This style was especially used in Europe, but on the upstroke there was no work done, and hence half the time was lost. This and other difficulties led finally to the adoption of the circular saw whose continuous cut and high speed saved much time and greatly increased the output.

The circular saw was invented around the end of the 18th century to convert logs into lumber. The first insertable teeth for this saw were invented by W. Kendal, an American, in 1826. The 19th century saw the general application of **circular saw** as its continuous cut and high speed saved much time and greatly increased the output.

The **bandsaw** is an endless belt of steel, having teeth formed along one edge and traveling continuously around an upper and lower wheel. The idea of the bandsaw dates back to at least 1809, but it did not attain its prominence in woodworking machinery until the last quarter of the 19th century. That it did not find the general application at an earlier period was due to the difficulty of securely and accurately joining the ends of the band. Today it is in service in sizes from a delicate filament used for scroll sawing to an enormous steel belt 50 feet long and 12 inches wide, traveling over pulleys 8 feet in diameter, making 500 revolutions per minute and tearing its way through logs much too large for any circular saw, at the rate of nearly 2 miles a minute.

Wood, as an engineering material with natural origin, also has some undesirable properties. To address the inherent drawbacks of wood, various wood reconstruction or composition methods have been explored, and the products are engineered wood products, such as **wood-based panels**. The fast development of wood-based panel industry began in the 1930s when the thermosetting synthetic **resin** was applied to produce **plywood**. This technology made the rapid production of plywood possible and the plywood industry was greatly accelerated due to the needs of World War II. Other wood-based panels also had their commercial applications in the first half of the 20th century. A large fiberboard plant designed to use wood raw materials was built in Minnesota, the U.S. in 1916. The first hardboard production plant was built in 1926. The **defibrator** developed by the Swedish was adopted worldwide in the 1940s and is still popular today. Around 20 years later, the invention of a **refiner** that can prepare pulp with very low bulk density gave birth to **medium-density fiberboard** (MDF). The first MDF plant was built in New York in 1965. Another technology boosted during World War II was **particleboard** technology. By 1955 there were 20 particleboard plants in the U. S..

Another way to enhance the performance of wood is to modify wood by physical, thermal or chemical approaches. The beginning of the wood preservation treatment industry can be traced back to the late 1830s, when a method for the pressure impregnation of timber was developed. Water-based treatment systems employing arsenates, chromates, fluorides, and nitrophenols were firstly applied at the beginning of the 20th century. In 1933, the first **copper–chrome–arsenic (CCA)** wood preservative was developed and proved to be an excellent wood **preservative** that was used in increasing amounts throughout most of the 20th century. The concern about the leakage of CCA to the environment during the service of wood boosted many alternative wood modification methods. Wood thermal modification, for example, improves wood dimensional stability and biological durability by heating wood in an inert atmosphere at around 200 ℃. Since no chemicals are added during the treatment, it is considered an environmentally friendly method.

Along with the development of wood processing industry is the development of wood science and technology (WS&T). From the late 19th century, many international and domestic WS&T institutes and organizations were established to boost the scientific study and cooperation. The **International Union of Forest Research Organization (IUFRO)** was founded in 1892 in Germany as a non-governmental international network of forest scientists. The International Association of Wood Anatomist (IAWA) was founded in 1931 in France, and the American Society of Wood Engineering was in 1958, which later became the International Society of Wood Science and Technology (SWST). The university-level WS&T education began in the early 20th century. In 1929, the first undergraduate WS&T program was established in the U. S., following which, another 12 were founded between 1941 and 1951. Today, there are 13 universities in the U. S. offering WS&T-related programs with the focus expanded from wood to bio-based materials. In Europe, the first university reference book on wood technology entitled *Technologies des Holzes* was edited by Franz Kollmann in 1936, and Germany is still one of the leading countries in the field of WS&T with respect to both industrial application and academic study.

In recent years, more property-enhancing methods have been brought out continuously to make the wood harder, and more stable and durable. It can now be used to supply heat in a more efficient way. Its composite products have been used to build higher buildings. The modified wood can withstand cruel weather and serve longer. Continuous and extensive researches in WS&T are the main contributor to the progress. Today, in China alone, as much as 17 universities provide the WS&T program, which is the indication of great industrial demands for the expertise and the guarantee of its future development. Traditional material as it is, wood will definitely serve mankind till the future.

Words and Phrases

1. **fiber*** *n* [C]: the slender threadlike elements or groups of wood fibers or similar cellulosic material resulting from chemical or mechanical defiberization, or both, and sometimes referred to as fiber bundles. 纤维

2. **wood*** *n* [U, C]: the tissues of the stem, branches, and roots of a woody plant lying between the pith and cambium, serving for water conduction, mechanical strength, and food storage, and characterized by the presence of tracheids or vessels. 木材(只有表示种、类、各色各样等概念时才可作为可数名词)

3. **hectare**(*abbr.* ha) *n*: 公顷, 土地面积计算单位, 面积为 10 000 平方米

4. **plantation** *n* [C]: 人工林, 也称为 planted forest

5. **deforestation** *n* [U]: 森林砍伐

6. **afforestation** *n* [U]: 造林

7. **Forest Stewardship Council**: 森林管理委员会

8. **Programme for the Endorsement of Forest Certification**: 森林认证体系认可计划

9. **growing stock**: 立木蓄积量

10. **bamboo** *n* [C, U]: 竹

11. **Moso bamboo** (*Phyllostachys edulis*): 毛竹

12. **old growth***: timber in or from a mature, naturally established forest. If the trees have grown during most of their lives in active competition for sunlight, the bole is usually straight and relatively free of limbs. 天然林(木材)

13. **log*** *n* [C]: a section of the trunk of a tree usually referring to a length suitable for conversion to commercial products. 原木

14. **second growth***: timber that has grown after the removal, whether by cutting, fire, wind, or other agency, of all or a large part of the previous stand. Often limited to that growth following removal of old-growth timber. 次生林(木材)

15. **sawdust** *n* [C]: 锯屑

16. **lumber*** *n* [U]: the product of the sawmill and planning mill usually not further manufactured other than by sawing, resawing, passing lengthwise through a standard planning machine, crosscutting to length, and matching. 板材

17. **circular saw** *n* [C]: 圆锯

18. **bandsaw** *n* [C]: 带锯

19. **wood-based panel***: 人造板

20. **resin*** *n* [C, U]: (1) a solid, semisolid, or pseudo-solid organic material that has an indefinite and often high molecular weight, exhibits a tendency to flow when subjected to stress, usually has a softening or melting range, and usually fractures conchoidally. (2) liquid resin—an organic polymeric liquid which, when converted to its final state for use, becomes a resin. 树脂(胶黏剂)

21. **plywood*** *n* [U]: usually a crossbanded assembly made of layers of veneer or veneer in combination with a lumber core or other wood-based panel material joined with an adhesive. Plywood generally is constructed of an odd number of layers with a grain of adjacent layers perpendicular to one another. Outer layers and all odd-numbered layers generally have the grain direction oriented parallel to the long dimension of the panel. 胶合板

22. **medium-density fiberboard (MDF)***: a composite panel product composed primarily of cellulosic fibers in which the primary source of physical integrity is provided through the addition of a bonding system cured under heat and pressure. 中密度纤维板

23. **refiner/defibrator** *n* [C]: 热磨机

24. **particleboard*** *n* [C]: generic term for a composite panel primarily composed of cellulosic materials, generally in the form of discrete pieces or particles, as distinguished from fibers, bonded together with a bonding system, and that may contain additives. 刨花板

25. **copper chrome arsenic (CCA)**: 铜铬砷, 一种木材防腐剂

26. **preservative*** *n* [C, U]: a chemical substance which, when suitably applied to wood, makes the wood resistant to attack by fungi, insects, marine borers, or weather conditions. 木材防腐剂

27. **International Union of Forest Research Organization (IUFRO)**: 国际林业研究组织联盟

Notes

1. 重要术语和表述的标注

本书对重点词汇或表述的选择主要根据它们在专业研究或业务中的重要性。重点词汇或表述在正文中首次出现时以粗体字标识, 并在文后进行解释。一些使用频率很高的关键词在词汇表中以★号加注, 并附上该词汇的英文释义, 以便读者更准确地了解它的含义, 释义来源为相关专业术语标准(详见第 3 章)或字典。

在英文文献中对重要内容进行强调时, 可采用斜体、粗体、下划线或字母大写等方法, 对教材中重要术语或术语表词汇的强调常用斜体或粗体。本书中斜体在正文中用于表示引文的书名或刊名, 重点词汇和表述则采用粗体标注。

2. 标题的书写原则

英文标题的书写除了应符合语法规范之外, 还有一些通用规则, 主要包括:

1) 标题中的主要词汇,包括名词、代词、动词、形容词、副词和一些连词的首字母用大写;
2) 冠词 the, a, an 用小写;
3) 介词多采用小写;
4) 常用的并列连词 and, but, for, or, nor 采用小写,如本章标题 Forest Resources and Wood Utilization;
5) to 无论用作介词还是作为动词不定式的一部分都采用小写,as 在任何语法功能下都采用小写,如本章中的小标题 Wood as a Sustainable Biomaterial。

3. 度量衡

本章涉及大量物理量的表达。在多数情况下,本书采用国际单位制 (Le Système international d'unités, 缩写 SI)单位,这是目前采用最为普遍的标准度量衡单位系统,旧称米制/公制(metric system)。公制的基本单位有 7 个,在英文文献中,这些单位常以缩写形式出现,分别为:

物理量	公制单位	缩写
length	meter	m
mass	kilogram	kg
time	second	s
electric current	ampere	A
thermodynamic temperature	kelvin	K
amount of substance	mole	mol
luminous intensity	candela	cd

在采用公制表达物理量时需要注意,其采用的数学只能在 0.1 ~ 1 000 之间,如 24 000 m 应表达为 24 km。采用缩写形式的单位没有单复数之分,但如果采用完整拼写则有单复数的差异,如 5 m=5 meters。

在木材贸易领域,木材的体积一般采用公制的衍生单位立方米(m^3)计算,但是美国采用的单位制为 United States customary units, 它源于英制单位(English units)。长度常用 inch(in), foot(ft), 质量采用 ounce(oz), pound(lb), 温度采用 degree Fahrenheit(℉)。因而美国的材积计算采用的是美制单位系统的衍生单位,常用的材积单位有:

1) board foot(板英尺): 1 板英尺表示 1 英尺长、1 英尺宽、1 英寸厚的板材的体积,缩写为 ft. b. m., bd. ft. 或 fmb。在实践中,由于板英尺是一个很小的单位,因而常以千板尺(1 000 board feet)来替代它,缩写为 M bd. ft., M B. M. 或 M B. F.。1 千板尺等于 2.360 m^3。

2) cord(考得): 加拿大和美国用于测量薪材或纸浆材体积单位。1 考得木材体积为 128 cubic feet (3.62 m^3)。

3) 另外需要注意的是,美国的文献和应用中在表达未刨光的粗锯材厚度时常以 1/4 英寸为基数进行表达,称为"1/4 系统"(quarter system),如 1 英寸厚板材常表达为 4/4(four quarters),而 5/4 则表示厚度为 1-1/4 英寸。这一系统的起源有不同的说法,一说是由于原木在锯解时,在带锯跑车上每次可偏移的最小位移档位为 1/4 英寸;也有认为这是由于粗材在刨光时所损失的厚度约为 1/4 英寸。

4. 连字符(hyphen)

连字符(-)是科技英语中的常用符号,是否需要使用连字符往往取决于使用长期使用过程中的约定成俗,而非明确的规则,因而较为复杂,本章中出现的连字符可以归纳为 3 类用法:

1) 复合名词,如 by-products。并不是所有的复合名词都需要用连字符连接,如 cutting tool, 有些复合名词在长时间使用后甚至最终会合成一个单词,如 sawmill, 因此是否要在复合名词的用连字符比较稳妥的做法是在字典中进行核查。如果不确定两个词是否可以用连字符相连,保守的做法是仍保留他们分隔的状态。

2) 复合形容词，用于所修饰的名词之前，这也是连字符最广泛的应用，如 4.0-billion-hectare forest, state-owned forest, wood-based panels, lumber-manufacturing equipment, large-scale commercial agriculture, old-growth forests。连字符用于复合形容词的一个主要功能是避免语义上的混淆，试比较：

- a small log processing system 小型原木加工系统
- a small-log processing system 小径原木加工系统

对于某些复杂的名词修饰，甚至需要多个连字符才可能理解所需表达的含义，如：

- urea-bonded medium-density mat-formed particleboard 脲胶黏合的中等密度铺装式刨花板

如果两个复合形容词的后半部分相同，则可以把第一个省略，如：

- high- or low-pressure condition 高压或低压条件
- PF- or MF-bonded boards 酚醛或三聚氰胺树脂胶合的板材

此外，在书写上需要注意，出现在复合形容词中的名词只采用单数形态，如：

- 4.0-billion-hectare forest 40亿公顷森林
- 10-day-long experiment 历时10天的试验

当然，如果在名词之前过多采作带连字符的复合形容词也会造成句子结构上的失衡，在这种情况下调整句子的结构是更好的选择，如：

Products from the hotter-than-expected full-sized vessel couldn't meet the technical standards.

可更改为：

Products from the full-sized vessel, which is hotter than expected, couldn't meet the technical standards.

3) 前缀，anti-, co-, ex-, mid-, non-, pre-, post-, pro-, self- 等，用以和后面的部分隔开，如 non-wood, non-governmental。

Exercises

1. Make a brief introduction to global forest resources and the recent trend.
2. Summarize China's forest resources and point out the characteristics of it.
3. Why wood is called a sustainable material? How to guarantee the sustainability of wood supply?

References

1. Chinese Forestry Society, National Poplar Commission. Forest Resource, Timber Production and Poplar Culture in China. In: Proceedings of International Conference on the future of poplar, Rome, Italy, November 13-15, 2003.
2. Food and Agriculture Organization. Global forest resources assessment 2015. Rome: FAO, 2016.
3. Food and Agriculture Organization. States of the world's forests 2016. Forests and agriculture: land-use challenges and opportunities. Rome: FAO, 2016.
4. Forest Product Laboratory. Wood handbook, Wood as an engineering material. General technical report FPL-GTR-190, 2010.
5. Hill C A S. Wood Modification Chemical, Thermal and Other Processes. West Sussex: John Wiley & Sons Ltd, 2006.
6. Zeng W, Tomppo E, Healey S P, et al. The national forest inventory in China: history-results-international context. Forest Ecosystems, 2015, 2:23.

Wood Structure and Chemistry

To pursue better use of wood, it is necessary to have some basic knowledge of the properties of this natural raw material. However, the properties of wood and its behavior significantly depend on the macroscopic and microscopic structures as well as on the chemical composition. This chapter starts with the origin of wood, followed by the macroscopic, microscopic structures and finally ends with the chemical composition and the submicroscopic structure of the wood cell wall.

Classification of woody plants

Trees are complex organisms. Wood is formed by a variety of plants, including many that do not attain tree stature. A tree is generally defined as a **woody** plant with 4 ~ 6 m or more in height and characterized by a single trunk rather than several stems. Plants of smaller size are called shrubs or bushes. Species that normally grow to tree size may occasionally develop as shrubs, especially where growth conditions are adverse. Because of the size attained, wood produced by plants of tree stature is useful for a wider range of products than wood from shrubs and bushes. For this reason, wood produced by trees is emphasized.

Woods are usually divided into two categories, namely, **softwoods** and **hardwoods**. Softwood and hardwood trees are **botanically** quite different. To define them botanically, softwood comes from gymnosperms (mostly conifers), and hardwood comes from angiosperms (flowering plants).

The typical characteristic of angiosperms is its seeds within ovaries, but gymnosperms produce seeds that without ovaries. Commonly, softwood trees are needlelike leaves and known as evergreen trees and are therefore often referred to as **conifers**. In the cold temperate zone of the Northern Hemisphere, softwood species include *Pinus*(pine), *Picea*(spruce), *Larix* (larch), *Abies*(fir), *Tsuga* (hemlock), *Taxodium* (cypress) and *Pseudotsuga* (Douglas fir), etc.

Unlike softwoods, hardwoods are typically broadleaves, that generally change color and drop in the autumn in temperate zones. The hardwood principally belongs to species that fall within the dicotyledon class, such as *Quercus* (oak), *Fraxinus* (ash), *Ulmus* (elm), *Acer* (maple), *Betula*(birch), *Fagus* (beech), and *Populus* (cottonwood, aspen) in the Northern Hemisphere.

In China, there are about 1 000 tree species that can be used as wood but only 300 in that are used commercially. Furthermore, the hardwood species are far more and widely distributed than softwoods. Today, the commercially important species of softwood are from Coniferopsida and hardwood are from Lauraceae, Juglandaceae, Fagaceae, and Salicaceae.

Generally, formal wood names are assigned by scientific Latin names, which are used in plant taxonomy. This scientific name system was founded by Carolus Linnaeus, a famous scientist in plant science from Sweden. Also, it can be called as terminology, meaning all the species names consist of two

Latin or Latinization words. The former word is the generic name, and the latter is the specific name, such as *Robinia pseudoacacia* L. In addition, the wood name also could show as generic name + sp. because of the similarly of wood properties in the same genus.

Macroscopic character of wood

A tree is a living organism, which is a course of life going through seed germination, seedling growth, flourishing branches, and mature trunk. The whole tree consists of the tree crown, trunk, and roots (5%~25% of tree total volume) (Fig.2.1). Although the crown plays an important role in gas exchange and photosynthesis, it also provides branches that can produce wood products, normally 5%~25% of total volume, for manufacturing wood-based panels. As the biggest part, trunk connects crown and root, which is the primary resources of wood, in the 50%~90% tree volume range.

Bark, xylem and pith, is distinguished under macroscopic, forming the basic structure of trunk (Fig.2.1). The whole tissue covering outside of **xylem** is bark, which is performed the function on protection and conduction. The bark is composed of an inner layer (**phloem**) and an outer protective layer (outer bark). Nearby the phloem, a thin and indiscernible layer, between the xylem and phloem produces new xylem and phloem tissue, is termed to be **cambium**, and also is called lateral meristem. In a growth season, new xylem cells are produced in the direction near pith and new phloem cells are produced in the opposite by cambium. Based on this view, cambium is the source of phloem and xylem. The tissue between cambium and pith is xylem, which is the most important part of the trunk. The xylem can be divided into primary xylem and secondary xylem in line with cellular source. The former holds the tiny part in the trunk and combines with pith. Moreover, the latter is the main body of the trunk, which is produced from cambium yearly. This part is the secondary xylem, which is the principal object in wood processing and utilization. The pith site in the center of the trunk is composed by softly parenchymatous tissue that makes wood quality lower. Wood consists of many cells, which are different in morphology, size, and arrangement. Cell tissue is an aggregation by all the cells that have the same function. A number of features in wood, such as growth ring, sapwood and heartwood, early wood and

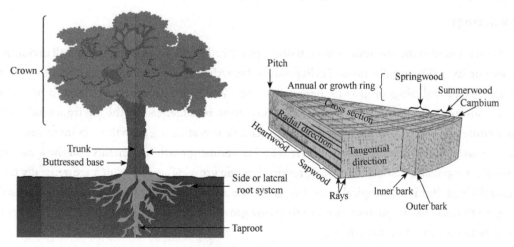

Fig.2.1 Parts of tree and macrostructure of wood

latewood, can be visually inspected and are termed to be macroscopic characters. Therefore, macroscopic characteristics of wood in fact is representation of cell or tissue under naked eye or low magnifiers. The macroscopic features provide clues of wood growth, the indication of physical properties, and assistance in wood identification and utilization.

Three distinct surfaces of wood

The appearance of sawn wood varies greatly according to the direction in which it is cut. These differences are not only in appearance but in physical properties as well. Generally, there are three planes termed as cross-sectional, radial and tangential surfaces which cut from a round cross-section. The cross-sectional surface can also be expressed as **transverse surface**, which is perpendicular to the longitudinal direction of the tree stem. Both **radial and tangential surfaces** are cut along the longitudinal direction. The radial surface needs to be cut through the pith, but the tangential surface does not (Fig.2.1). Above all, only the one cut that is in a plane through the pith can produce a truly radial surface. Intermediate planes of cut are commonly referred to radial to tangential according to which they more closely approximately.

Heartwood and sapwood

On cross section of xylem, the part closest to the bark is termed as **sapwood** (Fig.2.1). Sapwood is generally lighter in color because it consists of more live parenchymal cells and performs the function of water and mineral substance conduction in tree physiology. Being different from sapwood, the **heartwood** is regarded as the part from the pith to sapwood (Fig.2.1). Commonly, its proportion is higher than sapwood and has darker color resulted from more extractives. Due to the better wood properties in heartwood than sapwood, the formation mechanism has been investigated. Majority reports consider that heartwood is transformed from sapwood after the cells' inactivity with radial growth and wood cell senescence in the sapwood. There is a transition zone between sapwood and heartwood. With the heartwood formation, protoplasm in live cells is degraded and some extractives, such as tannin, resin, and pigment are deposited, resulting in that the water conduction system is jammed, which causes wood density increase and penetrability decrease.

Growth rings

As mentioned in the previous section, wood is produced by the vascular cambium cell divisions one layer by one layer. These collections of cells produced together over a continuous or discontinuous time are known as growth rings or growth increments (Fig.2.2). On the cross-section, growth rings always present concentric circle around the pith and its formation is influenced by the environmental factors. Temperature and precipitation are the most important factors affecting growth rings increment. In the cold temperate zone, the trees produce their wood in annual growth increments and show as distinct and regular rings that means all the wood produced in one growing season, which are generally termed as **annual rings**. But in the tropical zone, tree growth is always affected by precipitation and trees form their growth traces inconspicuous or more than one growth ring each year, so that many sources refer to as growth increment or growth ring.

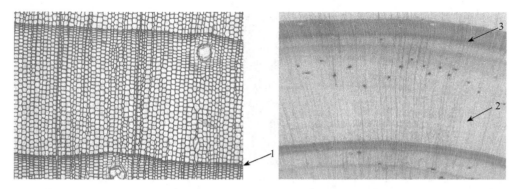

Fig.2.2 Growth ring (1), earlywood (2) and latewood (3)

Earlywood and latewood

As well known, wood is produced from cambium cell activities, which is significantly influenced by environmental factors, such as temperature and rainfall. For the tree growth in the cold temperate zone, in its early-stage of wood formation in a growing season, cambium cells are split rapidly and produce big-diameter cells, thin cell wall and lighter color wood, which is termed as earlywood (Fig.2.2). Corresponding to that, the latewood is often formed in the later stage of a year and appears small-diameter cells with thick walls and darker color. Due to the period of wood formation in a growing season, **earlywood** and **latewood** are also called as **springwood** and **summerwood**, respectively.

Because of the difference in wood cells, it is known that wood properties of earlywood are significantly different from that of latewood in distinct-ring softwood and hardwood. Many studies found that earlywood and latewood have different mechanical properties and different relationships between mechanical properties and other wood characteristics. Commonly, the strength and stiffness of latewood are obviously greater than those of earlywood. Therefore, the proportion of latewood in one growth ring is an important indicator of wood properties.

Pores

In fact, all cells in the wood are hollow structure and result in their **penetration**. As mentioned in the previous sections, the cell size and cell wall thickness are different according to cell types. The **Vessel** is regarded as hollow axial conducting tissue in hardwood trees and shows variously sized **pores** on the planes of the cross-section surface. The pores' diameters are larger than other cells and visible under macro-scale. In most cases, pores show rounded and separate from each other. Due to the vessel pores, hardwood is termed as porous wood. **Coniferous** wood and broad leave wood are different according to whether it has vessel pores. And the assembly, distribution, and arrangement of pores in the wood are closely related to the wood properties.

There are three types of pores based on its size and distribution in annual growth ring on the transverse section (Fig.2.3):

1) Diffuse-porous wood refers to which no difference of pores size in earlywood and latewood, such as *Populus* spp. , *Acer* spp. , *Betula* spp. ;

2) ring-porous wood, in which, the pore size in earlywood is much larger than that in latewood and annularity one or several rows with growth ring, such as *Quercus* spp. , *Ulmus* spp. , *Fraxinus* spp. ;

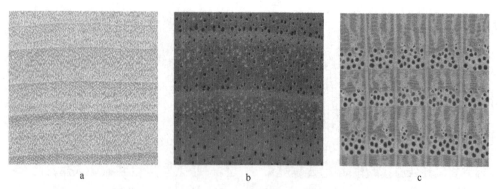

Fig.2.3 Three types of pores. a. *Populus trichocarpa* (diffuse-porous wood); b. *Juglans nigra* (semi-diffuse porous wood); c. *Quercus rubra* (ring-porous wood)

3) semi-diffuse porous or semi-ring porous wood, in which, the size of pores in earlywood is slightly larger than that in latewood and the pore diameter decreases gradually from earlywood to latewood in annual growth ring, such as *Cinnamomum* spp. , *Juglans* spp..

Rays

There are more than 90% cells with axial arrangement along with the tree height growth direction. However, another minority cell tissue radiation arrangement from pith to bark with shallow color is known as pith **rays**. The pith rays originate from primary tissue and extend outwards through cambium. The rays can be classified as xylem ray and phloem ray. Wood ray is a horizontal tissue and consists of ray parenchyma cell, playing an important role in transportation and storage of nutrient substances.

The wood ray in softwood is uniseriate and fusiform, so that is invisible under the macro scale. But the wood ray shows a significant difference in hardwood species, which is a key identification character. The same wood ray shows a various shape on three plane sections. On the transverse section, radiation shape is presented and the width and length could be measured. On the radial section, it shows linear or banding parallelism and the length and height could be measured. Generally, on the tangential section, it shows fusiform and the width and height are present. Overall, wood rays can be classified into tiny, medium and wide according to ray width under the macro scale.

Microscopic character of wood

Microscopic characteristics are wood structures observed via microscopic equipment. As well known, wood is composed of several different types of cells and the arrangement form results in various wood properties with different tree species. To investigate the wood structure and ultra-structure, it is vital to understand the factors affecting wood chemistry, physics, and mechanical properties. Based on these understandings, wood processing and utilization, as well as tree breeding can be carried out more effectively and successfully.

Softwood

Anatomy results show the simple and regular arrays of softwood cells, which are mainly composed of axial tracheid, wood rays, axial parenchyma and resin canal in softwood.

In softwood, all axial arrangements are slight and sclerenchyma cells, consisting of **tracheids**, resin tracheids and strand tracheids, which can be regarded as axial tracheids. The former one exists in whole softwood species but the latter two types only in tiny minority softwood. The tracheid cell is typically caecum, hollow, fusiform and contains many **pits** on its wall. Tracheid cells account for more than 90% in softwood and function as both water conduction and mechanical supporting tissue.

In Fig.2.4, it can be found that tracheid cells show radial arrangement on cross-section and the locations are staggered each other. The transverse magnification displays porous and difference on its morphology and cell wall thickness in earlywood and latewood. Because of different growing seasons, the cell wall is thicker and the cell lumen is larger in earlywood and inverse in latewood. Earlywood tracheid cell commonly is known as polygonal or hexagonal in earlywood but tetragonal in latewood. The average length of tracheid cells is 3 000 ~ 5 000 μm and the width is 15 ~ 80 μm. Tracheid length in latewood is always longer than that in earlywood. The thickness of the cell wall increases from earlywood to latewood and up to the maximum in the cells formed at the end of the growing season. As a result, softwood has an obvious growth ring boundary line. In one growth ring, the tracheid cell wall thickness changes gradually (*Pinus koraiensis*) or abruptly (*Larix gmelini*). Earlywood and latewood tracheid cells have the same tangential diameter and are related to the rough or smooth texture of softwood.

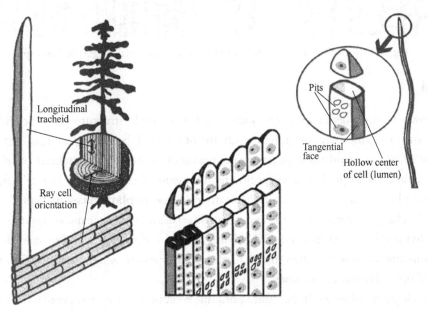

Fig.2.4 Microstructure of longitudinal tracheid in softwood (source: Shmulsky, 2011)

Wood ray is another principal cell in softwood but it is about 7% proportion of whole cells. All ray cells are composed of procumbent cells and every alone cell can be termed as ray cell. A great number of ray cells are parenchymal cell. The function of ray cells is different from that in sapwood and heartwood. Living, storage of nutrients and radial conduction are the features of rays in sapwood, while it shows dead and mechanical supporting in the heartwood. There are also sclerenchyma cells in some softwoods, such as *Pinus* spp. , *Picea asperata* and *Pinus taiwanensis*, being named as ray tracheid.

Uniseriate and fusiform wood rays are the types in softwood according to wood ray forms on the tangential section. Uniseriate wood rays are composed of only one-row cells or two rows occasionally, for instance, *Abies* spp. , *Cunninghamia lanceolata*, *Taxus chinensis*. Whereas some softwood species have radial resin canals in the center of multiseriate rays, which are termed as fusiform rays, such as *Pinus* spp. , *Picea* spp., *Larix* spp., etc.

Resin canal is a special **anatomy** structure in softwood and a duct surround by many thin-walled secretory cells. Its proportion is about 0.1% ~ 0.7% of the whole wood volume. There are normal resin canals (NRC) and traumatic resin canals (TRC) in softwood (Fig.2.5). The former one often exists in wood from following six genera, i. e. , *Pinus, Picea, Larix, Pseudotsuga, Cathaya* and *Keteleeria*. But the traumatic resin canal can be found in all softwood species after the wood formation in the tree was disturbed by external factors. NRC is always separated and randomly distributes in latewood. However, axial TRC distributes tangentially in earlywood on the cross section and normally near the growth ring.

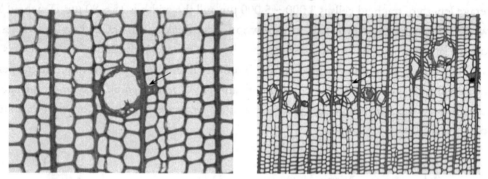

Fig.2.5　Normal resin canals (left) and traumatic resin canals (right) in *Pinus* spp.

Hardwood

As known in the previous section, softwood has a simple structure and tracheid plays both conduction and mechanical support function. On the other hand, besides wood from *Tetracentron* sp. and *Trochodendron* sp. , hardwood species are composed of markedly different kinds of cells, which are much more efficiently adapted to meet specific requirements. From the respective of plant evolution, hardwood trees have an advanced position than softwood trees in **plant classification**. For hardwood, vessel cells take charge of axial conduction and xylem fiber cells mainly mechanical support. In additional, axial parenchyma cells, wood rays and tracheid also exist in various hardwood species. Because of the heterogeneous and various cell types, many hardwood species are widely used for furniture, panels, flooring, and other decorative purposes.

The development of vessels is the most significant evolutionary step compared to softwood xylem. In hardwood, vessels are the particular conducting pathways and composed of single elements which are thin-walled cells with more or less open-end plates. Due to the thin cell wall and large diameter of these vessels, they appear as holes in the transverse surface and are termed pores. The details of pores arrangement in different wood species were introduced in the previous section. Because of the much variability in distribution and size of vessels between hardwood species, offering a useful feature for **wood identification**.

The size and length of vessel elements are different based on wood species and position in a single tree. Diameter in the tangential direction of the vessel is always measured for evaluation. Major tree species show that the vessel diameter ranges from 25 μm to 400 μm. Many reports suggested that vessel in latewood is shorter than that in earlywood of **ring-porous wood**. The tree often has a longer vessel in rapid growth periods compared to the slow periods.

For the longitudinal connection of two vessel elements, more or less open-end plates exist in this area and the interlinked pores are termed to be perforated while the joint cell wall is perforated plates. The pattern of these perforation plates is either simple, scalariform or foramina (Fig.2.6). There are also many pits on the vessel cell wall for radial conduction between two cells. The pattern type of the perforation plates, as well as the pitting arrangement, are not changed within a given species, and both are useful for microscopic wood identification. Connections to fiber cells, longitudinal and ray parenchyma and other cell types also form typically pits.

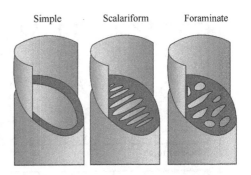

Fig.2.6 Perforate types in hardwoods

In the context of wood morphology, the term of fiber or fiber **tracheid** refers to another specific kind of cells. The long, tapered and usually thick-walled cells provide fibers with mechanical support for the hardwood xylem structure. Compared to softwood fibers, they are significantly shorter and tend to be rounded in the transverse section, while they show nearly rectangular shape in softwood tracheid. At the end of the annual growth, they become radially flattened. In hardwood, fibers and vessel elements are the two kinds of longitudinal cells commonly. The specialized vessels perform most conduction and the thick-walled fiber primary function is mechanical support. As a rule, a higher proportion of thick-walled fiber cells is invariably related to higher wood density and strength. Pitting between fibers and adjacent cells is either bordered or half-bordered pits, i.e., fiber-fiber pit pairs are normally bordered, whereas fiber-parenchyma pitting is typically half-bordered, pitting between fibers and vessels is rarely observed. A variation of the fiber, known as a *libriform fiber*, is marked by simple pits.

Axial **parenchyma**, thin-walled, tapered, short, brick-shaped epithelium around gum canals, is produced by cambium cell division. Concatenated longitudinally, parenchyma cells serve as food storage elements, which often cross walls and divide these longitudinal cells into a number of smaller sections during the process of cell maturation. In parenchyma cells, it probably contains oil, mucus or crystal depending on wood species and can be termed to be eleocyte, myxocyte or crystal idioblast. Whereas *axial parenchyma* is relatively rare in softwood species, it is often quite plentiful and significant in hardwood. The occurrence of axial parenchyma in hardwood is dependent on species and once more an additional evidence for wood identification. According to the adjacent relationship between parenchyma and vessels, parenchyma has been divided into apotracheal parenchyma and paratracheal parenchyma (Fig.2.7). In these two categories, several kinds were classified by the arrangement pattern of parenchyma cells on the transverse section. Due to the thin-walled cells, parenchyma tissue also relates to wood density and porous. More parenchyma often means low wood density and wood hardness.

The term of *wood rays* is regarded as parenchyma cells aggregation on the radial from pith to inner cambium. Besides minor species (*Populus* spp.), hardwood rays are much larger than softwood rays and the width in the tangential direction is up to about 30 cells. Moreover, wood rays account for up to 17% in hardwood volume. As mentioned in previous sections, softwood rays are generally uniserial but multi-seriate in hardwood (Fig.2.8). Also, unlike softwood, the cells of hardwood rays are all of the parenchyma type. Characterized by very large rays, hardwood, such as *Quercus* spp., exhibits distinctive ray patterns on both tangential and radial surfaces.

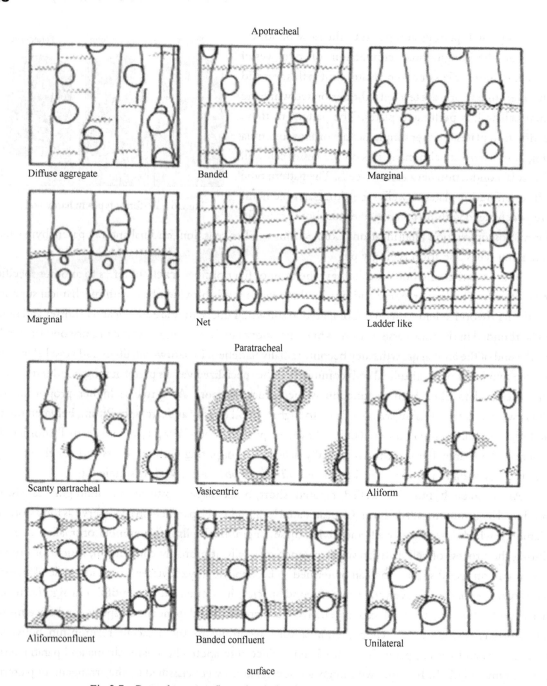

Fig.2.7 Parenchyma configuration in hardwoods on the transverse surface

The hardwood rays consist only of parenchyma cells, but they are different in cell shape and configuration. The ray parenchyma cells of hardwood are sometimes almost square when being viewed radially and have a rectangular shape. These cells are arranged perpendicular to the axial of longitudinal cells. Because these cells appear to be lying down, they are called to be *procumbent*. In some species, ray parenchyma on the upper and lower margins of the ray appears to stand upright on the end with its long axis parallel to the grain. These two kinds of ray configuration are termed as homogenous and heterogeneous rays, respectively. Homogenous rays consist of all cells procumbent, but heterogeneous rays entirely or partly consist of square cells or stand upright cells. The significance of ray cell configuration

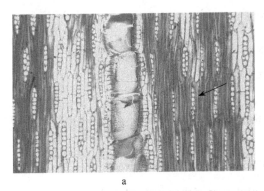 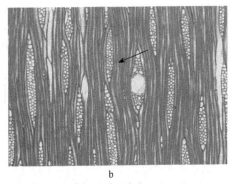

Fig.2.8 Wood rays in tangential section of *Pterocarpus santalinus* (a) and *Pistacia chinensis* (b); Arrows show uniserial ray (a) and multiseriate ray (b)

is that this feature can be used in wood identification because upright and square ray cells occur as a constant feature in only some species. As a result, this arrangement of rays is primarily of interest for wood identification.

To sum up, the anatomy shows an obvious difference in softwood and hardwood. Softwood has a simple cell type and uniform wood structure. However, hardwood is much younger in the evolutionary perspective and produces specific cells or tissues to meet specific requirements of tree physiology. Major hardwood species have vessels, which are absent in soft wood species. The different wood anatomy suggests that hardwood has complex and varied wood properties compared to softwood.

Chemistry of wood

Wood is a typical organic **polymeric** material. The physical, chemical and biological properties of wood can be understood by referring to the polymeric chemical constituents. Wood is composed by a greater number of mature cells and the chemical constituents are comprehensive results from these cells. One mature cell is divided into two parts, cell wall, and cell lumen, as well as intercellular space. In the wood chemical analysis, it can be expressed as principal and trace chemical components. With very small variations between different wood species, the chemical components are formed by three main elements, namely, carbon, oxygen, and hydrogen. The contents of these three elements are about 49%, 44% and 6% of wood by dry weight, respectively. As a living tree, nitrogen and other inorganic elements are also essentially involved in the metabolism during **wood formation** and tree growth. A number of these elements form cell wall principal macromolecules, such as polysaccharides and lignin, which exist in all wood species. As trace substances, differences of ash and extractives can also be found with wood species. The general scheme in Fig.2.9 summarizes the molecular compounds of wood.

Cellulose, **hemicelluloses** and **lignin** are the principal molecular components of wood cell wall. Being different from the elementary composition of wood, the molecular components show significant variation depending on not only wood species but also different parts in a tree. Table 2.1 shows the macromolecular substances content diversity between hardwoods and softwoods.

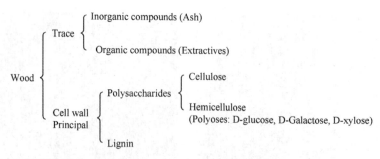

Fig.2.9 General category of molecular components of wood

Table 2.1 Macromolecular substances of the wood cell wall

Compounds	Content (%)	
	Softwoods	Hardwoods
Cellulose	42±2	45±2
Hemicelluloses	27±2	30±5
Lignin	28±3	20±4

Cellulose

The cellulose content of wood varies from about 40% to 50%. Cellulose is a homopolysaccharides and water insolubility. Based on glucose, cellulose forms larger linear chains by β (1→4) glycosidic linkages. The size of the linear polymer cellulose is expressed by the degree of polymerization (DP), which varies between 7 000 and 15 000 for woody celluloses. The polymer cellulose is held in larger aggregates and numerous cellulose chains are closely associated via intermolecular H-bonding to form **microfibril**. C1 of cellobiose in cellulose shows reducing end because it holds the hemiacetal molecular structure.

As well known, the microfibrils are key for enforcing wood cell wall and the microfibrils have **crystalline and amorphous** components with associated OH groups, which can be detected using a variety of methods, such as X-ray diffraction, nuclear magnetic resonance (NMR) and infrared (IR) spectroscopy. Because of the highly crystalline nature of the microfibrils, the cellulose component of wood is relatively unreactive and thermally stable. It appears that the crystalline region is associated with the core of the microfibril, with the exterior corresponding to the amorphous cellulose content.

Hemicelluloses

Another polysaccharide in wood is hemicelluloses, featured heteropolysaccharides, which may be highly branched, and are composed of several different monosaccharide molecules, such as galactose, arabinose, xylose, etc. As a result, hemicellulose has much lower DP (200 ~ 300) and molecular weight than that of cellulose. Generally, it is also less ordered and amorphous compared with cellulose. Depending on the basic saccharides, there are significant differences between softwoods and hardwoods, even individual wood species.

Like cellulose, hemicellulose has a number of hydroxyls in the saccharides, but contains more accessible OH, reacts more readily and is less thermally stable because of the general nature of hemicelluloses. Hemicelluloses appear to play a role as interfacial coupling agents between the high polarity surface of cellulose and the less polarity lignin matrix. Changes in structure or content of hemicellulose will affect the brittleness, viscoelastic and **hygroscopicity** properties of wood. It is found a generally higher proportion of hemicelluloses, a higher proportion of pentosans and a higher degree of acetylation in hardwood compared with softwood.

Lignin

The lignification processing in cell wall formation produced lignin, which is the obvious distinction between woody and herbaceous plants. Therefore, lignin is responsible for the compressive strength and stiffness of the wood cell wall. Lignin is a highly amorphous phenolic polymer and formed by three basic monomers, namely, p-coumaryl alcohol, coniferyl alcohol and sinapyl alcohol. These monomer units are polymerized through a free radical mechanism to produce a random three-dimensional network.

In wood cell walls, lignin joins together with cellulose and hemicellulose chains, filling up most of the micro voids between them. On the other hand, lignin also serves to bond individual cells together in the middle lamella region. Although lignin is relatively rigid at room temperature, it undergoes a glass transition at around 140 °C, and the presence of moisture in the cell wall additionally serves as a plasticizer for the lignin network. Softwood lignin consists mainly of phenyl propane units of the guaiacol type whereas both coniferyl and sinapyl alcohol present in hardwood lignin (Fig.2.10). Hardwood lignin has syringyl content varying from 20% to 60%, whereas softwood lignin has very low syringyl contents.

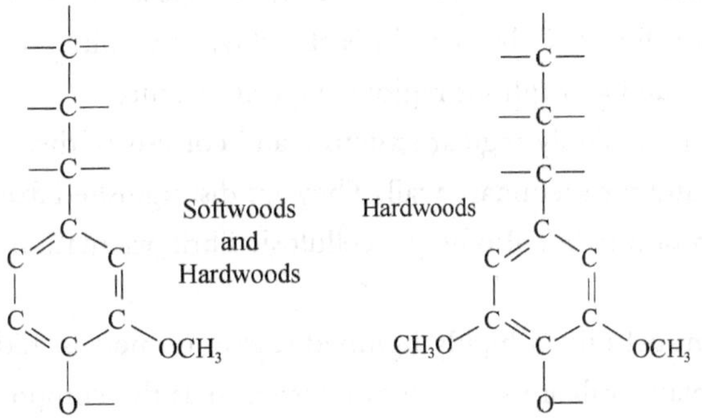

Fig.2.10 Basic monomers of lignin in softwoods and hardwoods

Extractives

The trace low molecular chemical components exist in the intercellular layer and cell lumen, being a non-structural role in wood. These substances extracted by organic solvents or water solution are termed to be extractives. All of these are products of secondary metabolites from photosynthesis in tree physiology. A great variety of wood extractives always depend on the solvent types, wood species, wood locations and the environmental factors on wood formation. Special components extracted from some woods are extensively used in the phytochemistry field. Wood extractive contents are 2% ~ 5% in wood dry weight and it consists of aliphatic series, terpenes, and aromatics.

Wood cell wall chemical structure and layers

The principal ingredients in the wood are cellulose, hemicelluloses and lignin, which have different responsibilities in the wood cell wall. The celluloses in aggregates, called elementary fibrils, are responsible for the tensile strength and skeleton to wood. Hemicelluloses permeate into a skeleton as well as amorphous polymer and enforce cell wall rigidity. Finally, lignin, the last formed polymer in the cell wall, combines all polysaccharides and gives cell wall stiffness (Fig.2.11).

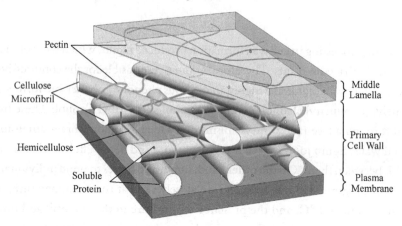

Fig.2.11　Simulation structure of polymers in wood cell wall

Cellulose, the skeleton of the wood cell wall, basically formed by elementary fibrils, consists of approximately 36 ~ 42 parallel cellulose chains; microfibrils are composed of several elementary fibrils shrouded by shorter hemicellulose chains. The larger morphological units termed as fibrils are composed of bundles of cellulose microfibrils. Cellulose fibrils also show crystalline regions with highly ordered parallel chain arrangements and amorphous regions opposite features.

The wooden cell wall is a highly regular structure and consists of three main regions: the middle lamella, the primary wall, and the secondary wall. They are distinguished from each other, not only by the macromolecular composition, but also by the cellulosic fibril orientation to the longitudinal axis of the cell.

Between two adjacent cells lies a highly lignified region termed as middle lamella. Both middle lamella and adjoining primary wall are sometimes referred to as the compound middle lamella. After middle lamella formed, primary wall deposited in the inner of the middle lamella. Due to the enlargement and elongation of the cell, the primary wall is commonly thin and resilient. Primary wall basically consists of largely random orientation of cellulose microfibrils with an angle from 0 to 90 degrees relative to the long axis of the cell.

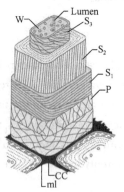

Fig.2.12　Simulation structure of wood cell wall layers and microfibril angle (source: Agarwal, 2006)

Fig.2.12 shows structural details of a layer in the wood cell wall. After the stop of cell enlargement, the secondary cell wall is formed in the direction to the cell lumen and subdivided into three layers, i.e., S_1, S_2, and S_3. S_1 is adjacent to compound middle lamella and is a thin layer which is characterized by a large microfibril angle of 50° ~ 70° degrees relative to the long axis of the cell. The next formed S_2 is the most important layer related to the mechanical properties of wood. This layer is the thickest secondary cell wall layer. S_2 is featured by a lower lignin percentage and a low microfibril angle of 5° ~ 30° degrees. Interior of the S_2 layer is the S_3 layer, another relatively thin wall layer. Because of it is adjoining to water transportation in living wood cells, low lignin concentration and high microfibril angle degrees (60° ~ 90°) in the S_3 layer.

Pits in the wood cell wall are connections in adjoining cells. These features directly influence wood processing and utilization, which can be used in wood identification. Parenchyma cells form thin secondary walls, but others form thicker secondary walls. In all wood cell types, the secondary wall layers show specific openings or gaps, being termed as pits. These pits allow substances to flow between adjacent cells. This role is not performed in living trees but also affects the quality of wood drying, permeation, and chemistry modification. Simple and bordered pits are the two main types and the structure and shape of pits varies with different wood cells. Simple pits appear in connecting two parenchyma cells and bordered pits exist in connecting two prosenchyma cells (Fig.2.13).

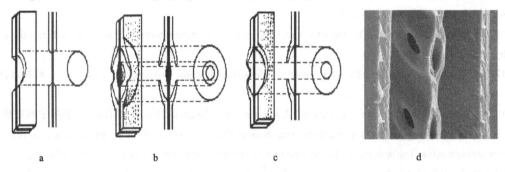

Fig.2.13 Pit types. a. Simple pits; b. Bordered pits; c. Half-bordered pits; d. SEM shows pits in *Pinus koraiensis* (source: Liu, 2013)

Words and Phrases

1. **woody** *adj*: a generic term applied to a group of materials manufactured from wood or other lignocellulosic fibers or particles to which binding agents and other materials may be added during manufacture to obtain or improve certain properties. 木质的或木基的，在植物原料或材料中常表示为木质的，而在复合材料中常用木基的

2. **softwood*** *n* [C, U]: generally, one of the botanical groups of trees that in most cases have needlelike or scale-like leaves; the conifers; also the wood produced by such trees. The term has no reference to the actual hardness of the wood. 针叶树材或软材(与木材硬度无关)

3. **hardwood*** *n* [C, U]: generally one of the botanical groups of trees that have broadleaves in contrast to the conifers or softwoods. The term has no reference to the actual hardness of the wood. 阔叶树材或硬材(与木材硬度无关)

4. **botanically** *adv*. 植物学上

5. **conifer** *n* [C]: 针叶树, 形容词形式为 coniferous

6. **bark*** *n* [U]: the layer of a tree outside the cambium comprising the inner bark and the outer bark. 树皮, 通常可以分为内皮和外皮。内皮包含大量生活细胞，在树木生长中传输光合产物。外皮多为死细胞，主要由软木脂、单宁等组成，起保护的作用

7. **xylem** *n* [U]: the portion of the tree trunk, branches, and roots that lies between the pith and the cambium. 木质部, 木材利用的主要部分

8. **phloem** *n* [U]: inner bark; the principal tissue concerned with the translocation of elaborate foodstuffs. 韧皮部, 树木生长中输导光合产物, 与 bast 同义

9. **cambium** *n* [U]: the layer of cells that lies between the inner bark and the wood of a tree, that repeatedly subdivides to form new wood and bark cells. 形成层，是侧生分生组织，是木材直径生长的基础。向内分生出木质部母细胞，

向外分生出韧皮部母细胞

    10. **transverse surface:** 横截面

    11. **radial surface:** 径面

    12. **tangential surface:** 弦面

    13. **sapwood*** *n* [C]: the wood containing some living cells and forming the initial wood layer beneath the bark of the log. The thickness of the sapwood layer varies by species and may be lighter in color than heartwood. Under most conditions, the sapwood is more susceptible to decay than heartwood. 边材

    14. **heartwood*** *n* [C]: the inner layer of a woody stem wholly composed of nonliving cells and usually differentiated from the outer enveloping layer (sapwood) by its darker color. It is usually more decay resistant than sapwood. 心材

    15. **annual ring*** *n* [C]: the growth layer produced by the tree in a single growth year, including earlywood and latewood. 年轮, 常出现在寒温带树木中

    16. **earlywood/springwood** *n* [U]: the less dense, large-celled, part of the growth layer formed first during the annual growth cycle. 早材或春材

    17. **latewood/summerwood** *n* [U]: the denser, smaller-celled, later-formed part of a growth layer. 晚材或秋材。同 latewood

    18. **penetration** *n* [U]: the entering of chemicals into the wood. 渗透或浸渍, 是木材改性与利用中的重要工艺

    19. **vessel*** *n* [C]: wood cells of comparatively large diameter that have open ends and are set one above the other so as to form continuous tubes. The openings of the vessels on the surface of a piece of wood are usually referred to as pores. 导管

    20. **pore*** *n* [C]: in wood anatomy, a term applied to the cross-section of a vessel or of a vascular tracheid. 管孔

    21. **ray** *n* [C]: 木射线

    22. **tracheid** *n* [C]: 管胞

    23. **pit*** *n* [C]: 纹孔

    24. **anatomy** *n* [U]: the scientific study of the structure of an animals or plants, or of a particular type of animal or plant. 解剖或解剖学

    25. **plant classification**: the act or process of dividing plants into groups according to their genetic relationship and evolution. 植物分类, 通常是基于植物的进化程度

    26. **wood identification**: the act of recognizing and naming wood genus or species. 木材鉴定, 依据木材宏观和微观特征确定木材名称的过程

    27. **ring-porous wood***: hardwoods in which the pores of the earlywood are large compared to the latewood, thus forming a distinct zone or ring of pores. 环孔材

    28. **diffuse-porous wood***: certain hardwoods in which the pores exhibit little or no variation in size or distribution throughout the growth ring, only decreasing slightly in size, gradually toward the outer border of the ring. 散孔材

    29. **parenchyma** *n* [C]: 薄壁细胞

    30. **polymeric** a material formed by the chemical reaction of molecules to form higher molecular weight molecules consisting of repeating units. 聚合物的、聚合的

    31. **wood formation**: the process of wood produced by cambium. 木材形成, 包括了形成层细胞分化、细胞增大与伸长、细胞壁增厚、木质化、细胞程序性死亡及心材形成等过程

    32. **cellulose*** *n* [U]: the carbohydrate that is the principal constituent of wood and forms the structural framework of the wood cells. 纤维素, 以 β-1,4 糖苷键连接的 D-葡萄糖残基组成的链状高分子化合物, 是非常重要的自然资源

    33. **hemicellulose*** *n* [U]: the carbohydrate that consists of two or more glycosyl and serving as a binder in the wood cell wall. 半纤维素

34. **lignin*** *n* [U]: a substance found in the edges of some plant cells that makes the plant hard like wood. 木质素，主要增强木材细胞壁。在木材细胞壁形成中，木质素的生物合成与沉积过程被称之为木质化 Lignification

35. **microfibril*** *n* [U]: a structure consisting of approximately 40 cellulose molecular chains. 微纤丝，纤维素的基本组成单位。细胞壁中微纤丝排列的角度与木材性质密切相关

36. **crystalline and amorphous*** *adj*: the spatial structure of cellulose. Crystalline means having the regular structure like a crystal, but amorphous shows disorderly and unsystematic. 纤维素结晶区与无定型区，两者之间无明显的绝对界限

37. **hygroscopicity** *n* [U]: 吸湿性

Notes

木材分类与命名

木材产自与树木，它的分类主要依照树木分类，有界、门、纲、目、科、属、种。而在常用的木材中，科、属、种最为重要。树木分类中是根据花、果、叶等形态特征进行分类，但是在木材中有的时候难以或不需进行精确分类，所以木材分类中以属为基础、材性为依据，进行归类。

汉语	拉丁文	词尾	英语	举例	
界	Regnum		Kingdom	植物界	Plantae
门	Divisio(Phylum)	-phyta	Phylum	裸子植物门	Gymnospermae
纲	Classis	-opsida, -eae	Class	松杉纲	Conifer-opsida
目	Ordo	-ales	Order	松杉目	Pin-ales
科	Familia	-aceae	Family	松科	Pin-aceae
属	Genus	-us, -a, -um	Genus	松属	*Pinus*
种	Species		Species	红松	*koraiensis*

木材的名称分为俗名和学名两种。俗名是人们用地方语对该木材的命名。学名是利用拉丁文或拉丁化的语言对该木材进行命名的，是世界通用的木材名称，它的命名法则由《国际藻类、真菌、植物命名法规》(*International Code of Nomenclature for algae, fungi, and plants*)规定。仅用俗名无法准确指称木材所属的类别，有时很多不同的俗名其实指的是同一材种，如 African ebony(非洲乌木)，在不同国家的称谓并不相同，也称 Msindi (坦桑尼亚), Omenowa (加纳), Nyareti (尼日利亚)或 Kukuo (冈比亚)，它们都表示柿属的一类木材 *Diospyros* spp. 。

在英文文献中很多木材的英文俗名不能简单直译成中文，否则也会与实际情况形成很大偏差，常见的例子有：

拉丁名	英文名	中文名	误译名	误译名对应准确英文名	误译名对应准确拉丁名
Liriodendron tulipifera	yellow poplar	北美鹅掌楸	黄杨	Korean boxwood	*Buxus sinica*
Pinus resinosa	red pine	美国赤松	红松	Korean pine	*Pinus koraiensis*
Sequoioideae	redwood	北美红杉	红木	hongmu	
Pseudotsuga menziesii	Douglas fir	北美黄杉	花旗松*	—	—
Pinus sylvestris	Scots pine	欧洲赤松	苏格兰松	—	—
Picea abies	Norway spruce	欧洲云杉	挪威云杉	—	—

*花旗松虽然是一种误译名，但因被长期使用目前也被普遍接受。

木材树种的科名是一个复数形容词作名词用，有词干和词尾(aceae)两部分组成。大多数科的词干是该科

命名时的模式属的属名。如杉科 Taxodiaceae, 词干来自模式属落羽杉属 *Taxodium*; 杨柳科 Salicaceae, 词干来自模式属柳属 *Salix*。

木材的属名来源广泛, 如源自古希腊文的桑属 *Morus* (希腊语中桑为 Morea); 来自经典拉丁文的槭属 *Acer* (拉丁语中槭树为 acer); 以古代神话命名的胡桃属 *Juglans* (Jove 是古罗马主神的名字+glans 橡子); 纪念某重要人物的黄檀属 *Dalbergia* (Nicholas Dalberg, 1736—1820, 瑞典植物学家); 或含有特殊化合物成分的云杉属 *Picea* (拉丁语中的松脂 picea)。

木材的学名参考了植物分类中树木的名称, 其结构为属名+种名+定名人。在书写时属名第一个字母大写, 其他字母小写; 种名字母均为小写, 定名人常用省略词表示, 第一个字母大写。属名和种名需要斜体, 如: 欧洲赤松的学名为 *Pinus sylvestris* L., 表示它属于松属 *Pinus*, 种名为 *sylvestris*, 命名人为 Linnaeus(林奈)。如果在一篇文章中多次提到某个树种, 除第一次提及时给出全写, 在以后出现时可将属名缩写, 但绝不能省略, 如上例可简写为 *P. sylvestris*。当某个树种无法确定到种时, 可只写属名, 后面加以 sp. (单数) 或 spp. (复数), 如: *Populus* sp.。

Exercises

1. Summarize the differences between anatomical and chemical properties between softwood and hardwood.
2. How to comprehend the microstructure of wood cell wall?
3. Think about the relationship between wood properties and its utilization.

References

1. Rowell R M. Handbook of Wood Chemistry and Wood Composites. CRC Press, Boca Raton London New York Washington, D. C, 2005.
2. Shmulsky R, David J P. Forest Products & Wood Science: An Introduction (The 6th Edition). A John Wiley & Sons, Inc., Publication, 2011.
3. Callum A S. Hill. Wood Modification, Chemical, Thermal and Other Processes. A John Wiley & Sons, Ltd. Wiley Series in Renewable Resources, 2006.
4. 刘一星, 赵广杰. 木材学. 2版. 北京: 中国林业出版社, 2013.
5. 沈显生. 植物学拉丁文. 合肥: 中国科学技术大学出版社, 2005.

Wood-Moisture Relations and Properties of Wood

Wood, like many natural materials, is hygroscopic, which takes moisture from the surrounding environment. **Moisture** exchange between wood and air depends on the relative humidity and temperature of the air and the current amount of water in the wood. This moisture relationship has an important influence on wood properties. Many challenges of using wood as an engineering material arise from the changes in moisture content in the wood during its service.

This chapter discusses the macroscopic physical properties of wood with the emphasis of the relationship with moisture content. Some properties are species-dependent; in such cases, data from the literature are tabulated according to species. The chapter begins with a broad overview of wood–water relations, defining key concepts needed for understanding the physical properties of wood.

Wood moisture and the environment

Measurement of wood moisture content

The moisture content (MC) of wood is the amount of water in wood as the percentage of its oven-dry weight. In equation form, MC is calculated by

$$\mathrm{MC} = \frac{m_w}{m_d} \times 100\% \tag{3-1}$$

where m_w and m_d is the mass of water and oven-dried wood, respectively. Operationally, the MC of a given piece of wood could be expressed by

$$\mathrm{MC} = \frac{m_i - m_d}{m_d} \times 100\% \tag{3-2}$$

where m_i is the mass of wood at a given moisture condition.

Based on the international standard (ISO 13061-1: 2014 Physical and mechanical properties of wood-Test methods for small clear wood specimens - Part 1: Determination of moisture content for physical and mechanical tests), the basic gravimetric method of measuring wood MC is depicted as follows: First of all, the test piece for MC determination should be prepared and stored under conditions which ensure that their MC remains unchanged. Weigh the test piece to an accuracy of 0.5% of its mass, m_i, before the drying procedure. Then the moist test piece is dried in an oven maintained at a temperature of 103°C±2°C until the reference weight m_d is attained. Based on the values of m_i and m_d, the MC of the test piece could be derived by Equation (3-2).

Equilibrium moisture content of wood

Wood in the living tree generally has an MC of 30% or greater, that is, the cell wall is fully saturated with water (Fig.3.1a). The cell cavity generally contains some water, the amount of which varies greatly among trees and among cells in the same tree. When the green wood is exposed to atmospheric conditions after the tree is felled, it loses moisture from the cell cavity initially. The MC, at which the cell cavity contains no liquid water, but the cell wall is fully saturated with moisture (Fig.3.1b), is called the **fiber saturation point (FSP)**. The FSP averages about 30% MC, but varies by several percentage points among tree species, and even between logs cut from different heights in a tree. When wood loses moisture further from cell wall to a sufficiently low MC to be at equilibrium with the ambient atmosphere (Fig.3.1c), this MC is designated as the equilibrium one, namely, **equilibrium moisture content (EMC)**, approximately proportional to the ambient **relative humidity (RH)**. For a given RH condition, EMC varies somewhat between heartwood and sapwood, with the type and proportions of cell wall constituents and extractives. It is also affected by temperature, heating history, and mechanical stress. Some or all these factors interact to some extent.

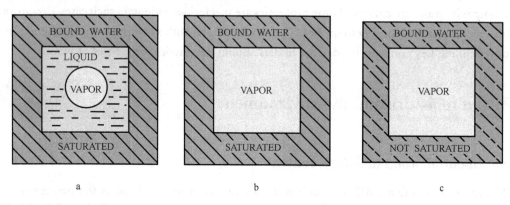

Fig.3.1 Schematic diagram showing an idealized representation of the moisture distribution in a wood cell cross-section. a. above, b. at and c. below FSP

The most important factor affecting the EMC of wood is the RH (or relative vapor pressure) around the wood. The curve that EMC changes as a function of RH is called as **sorption isotherm**. Like other hygroscopic polymers, such as cotton, wool, and cereal grain, wood exhibits a typically sigmoid isotherm as shown in Fig.3.2. As shown in Fig.3.2, at a given RH, the higher the temperature, the lower the EMC could be observed. The influence of temperature on wood EMC is ascribed as two aspects: 1) The immediate effect is to reduce the EMC at a given RH, which is a temporary reversible effect. 2) The permanent reduction in its hygroscopicity after the wood returns to room temperature. The extent of permanent reduction depends on the temperature and the exposing duration. To eliminate the temperature

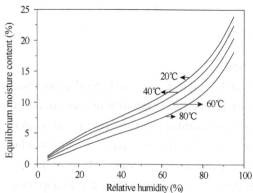

Fig.3.2 Typical sorption isotherms for Chinese fir at four temperatures

effect on moisture change in the wood, EMC values must be determined at a specific constant temperature.

Adsorption/desorption and sorption hysteresis

Generally, at a given RH, the wood EMC could be attained either in adsorption or desorption process. The attaining of equilibrium from adsorption/desorption process is an exponential function of time (Fig.3.3). It may require weeks or even months to attain the true equilibrium, being attributed to the slow molecular rearrangements during the destabilization of wood cell walls.

Fig.3.3 shows that the adsorption EMC curve lags below the desorption one, which is called **sorption hysteresis**. The desorption EMC is measured by placing a wet wood sample in dry condition. The adsorption EMC is measured in the opposite direction (from the dry state to a relative wet condition). The ratio of adsorption EMC to desorption EMC (A/D) could describe the magnitude of the sorption hysteresis effect and varies with wood species, RH, temperature, extractive content, etc. Some explanations have been applied in clarifying the mechanism of moisture sorption hysteresis, such as the effective hydroxyl group theory and the ink bottle effect.

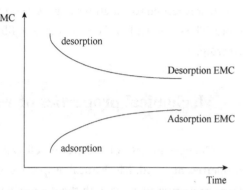

Fig.3.3 Adsorption and desorption curves approach equilibrium with increasing time

Hygro-expansion in wood

In the living tree, the cell walls of wood are completely saturated with **bound water**, and additional water may reside in lumens. At the living state or when MC is higher than FSP, wood is dimensionally stable. Below FSP, wood dimension changes as it gains or loses moisture because the amount of bound water is related to the volume of the wood cell wall. When the amount of bound water increases or decreases, cell wall **swells** or **shrinks**. Both the swelling and shrinking are considered under the general subject of "hygro-expansion".

Hygro-expansion in wood is traditionally given in terms of percent dimensional changes (percent shrinkage or percent swelling). The two terms differ with respect to the reference dimension used. For example, when the percent shrinkage is determined, the moist dimension is used as the reference base (l_1), and Eq.(3-3) can be used

$$\beta = \frac{l_1 - l_0}{l_1} \times 100\% \qquad (3\text{-}3)$$

where β is the percent shrinkage, and l_0 is the dry dimension after shrinkage.

Wood is **orthotropic** in the three primary directions, and therefore the hygro-expansion is orthotropic as well. Wood has the most shrinking (swelling) in the direction of the annual growth rings (tangentially, T), about half values in the direction of the across rings (radially, R), and very small values in the direction along the grain (longitudinally, L). From FSP to the completely dry condition, the percent shrinkages β in T, R, and L directions are 6% ~ 12%, 3% ~ 6% and 0.1% ~ 0.3%, respectively.

Generally, the hygro-expansion of hardwoods is greater than softwoods. In addition, species of high

specific gravity shrink or swell more than ones of low specific gravity. However, there are exceptions. Basswood, a light species, has high shrinkage, while the heavier black locust has more moderate shrinkage.

The considerable difference in the transverse hygro-expansion is responsible for **warping, checking,** and **splitting** of lumber associated with the initial drying process and during subsequent uses.

Commonly, the index "differential shrinkage" – the ratio of $β_T$ to $β_R$ – is applied to evaluate the anisotropic shrinkage and the stress extent during drying process.

The explanations of transverse anisotropy of hygro-expansion could be divided into three aspects. The first is related to variations in gross wood structure, such as arrangements of various tissues and cell types. The second is based on variations in fibrils alignment and the third relates to variations in cell-wall layering.

Mechanical properties of wood

Changes in MC of the wood cell walls below the FSP have a major effect on the mechanical properties of wood. Mechanical properties of wood are changed slightly at MCs above the FSP, while these properties increase with the decreasing MC significantly. This chapter includes discussion of the effect of MC on mechanical properties of wood. In Chapter 3.2, wood is described as an orthotropic material which exhibits varied hygro-expansion in three mutually perpendicular directions: L, R and T. Similarly, wood has different mechanical properties in the three directions as well. In addition, because wood is a natural and anisotropic material, the mechanical properties vary considerably. This chapter also provides information about the nature and magnitude of variability in mechanical properties.

Elastic and strength properties

As an orthotropic material, the elastic behavior of wood is described as twelve elastic constants (nine are independent), which are: three moduli of elasticity E, three moduli of rigidity G, and six Poisson's ratios $μ$. Elasticity implies that deformations produced by low stress are completely recoverable after loads are removed. Usually, the **modulus** of elasticity is determined from bending, E_L, rather than from axial tests (tension and compression), probably due to the importance of **modulus of elasticity** for structural applications.

Represented as "strength properties", mechanical properties are most commonly measures for design, including: 1) **Modulus of rupture** - Reflects the maximum load-carrying capacity of a member in bending and is proportional to maximum moment borne by the specimen. Modulus of rupture is an accepted criterion of strength, although it is not a true stress because the formula by which it computed is valid only to the elastic limit; 2) **Tensile strength** parallel to grain – Maximum tensile **stress** sustained in the direction parallel to grain; 3) Compressive stress perpendicular to grain – Reported as stress at proportional limit. There is no clearly defined ultimate stress for this property; and 4) **Hardness** – Generally defined as resistance to indentation using a modified Janka hardness test, measured by the load required to embed an 11.28 mm ball to one-half of its diameter. Values presented are the average of radial and tangential penetrations.

Influence of environmental factors on mechanical properties

Wood **stiffness** and strength are related to the amount of moisture in the wood fiber cell walls.

Above the FSP, water accumulates in the wood cell cavity and there are no tangible strength effects associated with a changing MC. When MC varies between 0 to the FSP, increasing amounts of bound water, and decreasing the stiffness and strength of wood. When water molecules penetrate wood cell walls, breaking hydrogen bonds between the polymers and forming hydrogen bonds between water molecular and amorphous components (hemicelluloses, lignin, and paracrystalline cellulose). The plasticization of the amorphous polymers enhances the flexibility of the polymer network. Since celluloses, hemicelluloses and lignin have different absorbability, the extents of hygro-expansion vary within the wood cell walls, providing the shear slip between cellulose and matrix, and leading to relatively large energy dissipation. However, not all mechanical properties change with MC, for example, the impact bending (height of drop causing complete failure) nearly does not vary no matter how MC changes.

When the effects of MC on stiffness and strength of wood are investigated, the temperature is usually fixed at room temperature (about 25 °C to 30 °C). However, wood mechanical properties are related to the temperature of the testing or working environment. The effect of temperature on mechanical properties is complicated, especially for moist wood. When the testing temperature is above 0 °C, both immediate and permanent effects should be considered, respectively. The immediate effects of the increased temperature are an increase in the plasticity of lignin and an increase in spatial size, which reduce intermolecular contact and are recoverable. The permanent effects manifest as an actual reduction in wood substance or weight loss via degradation mechanisms and are thereby non-recoverable. The effect of low temperature (<0 °C) on the stiffness and strength of wood is displayed in Fig.3.4. Stiffness and strength increase with the decreasing temperature. No matter MC higher or lower than the FSP, the higher the MC, the greater and faster the increase in stiffness and strength, resulted from the freezing free water in cell lumen and the formation of special icicle form in the low-temperature environment.

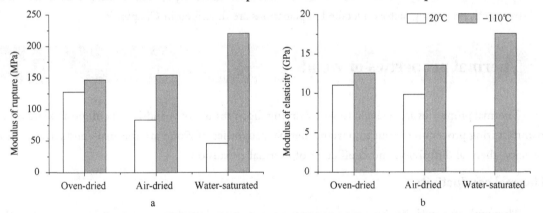

Fig.3.4 Changes of modulus of rupture (a) and modulus of elasticity (b) for birch wood at varied moisture states

Influence of natural characteristics on mechanical properties

Commonly, the straight-grained wood with no defect is used for assessing the stiffness and strength properties. However, the mechanical performance of wood varies in specific gravity, cross grain or knots due to natural growth characteristics. Natural defects such as **pitch pocket**s may occur as a result of biological or climatic elements influencing the living tree. These wood characteristics should be taken into account in assessing actual properties or estimating the actual performance of wood products.

The substance of which wood is composed is actually heavier than water; its specific gravity is about

1.5 regardless of wood species. In spite of this, dry wood of most species floats on water, which is thus evident that part of the volume of wood is occupied by cell cavities. Variations in size of these openings and in the thickness of the cell walls cause some species to have more wood substance per unit volume than other species and therefore higher specific gravity. Thus, the specific gravity is an excellent index of the amount of wood substance contained in a piece of wood; it is a good index of mechanical properties as long as the wood is clear and straight grained. However, specific gravity values also reflect the presence of **gums**, resins, and extractives, which contribute little to mechanical properties.

In some wood products, the directions of applied stress may not coincide with the natural axes of fiber orientation in the wood. Stiffness and strength properties in directions other than along the natural axes could be obtained from elastic theory or approximated equation when regarding wood as the two-phase composites, namely, microfibril and matrix (hemicelluloses and lignin).

Knot is the portion of a branch that has become incorporated in the trunk of a tree. The influence of a knot on the mechanical properties of a wood member is due to the interruption of continuity and change in the direction of wood fibers associated with the knot. Most mechanical properties are lower in the section containing knots than in the clear straight-grained wood because (a) the clear wood is displaced by the knot, (b) the fibers around the knot are distorted, resulting in cross grain, (c) the discontinuity often occurs around, leading to stress concentrations, and (d) checking often occurs around the knots during drying. However, transverse hardness and strength are exceptions.

Wood is usually treated with chemicals to enhance its fire-retardant or decay-resistance performances in services. Each set of treatment chemical and process has a unique effect on the mechanical properties of the treated wood. Commonly, the influences of **fire-retardant** or preservative treatments on mechanical properties of wood vary among treatment method, chemical type, and even pre-treatment and post-treatment processing factors. Detailed explanations are described in Chapter 7.

Thermal properties of wood

Thermal properties are relevant to wood drying, hot-pressing of wood-based composites and other manufacturing procedures. Four important thermal properties of wood are thermal conductivity, heat capacity, thermal diffusivity, and coefficient of thermal expansion.

Thermal conductivity

Thermal conductivity k is a measure of the rate of heat flow through a material subjected to the unit temperature difference across the unit thickness. k of common structural woods is much smaller than the thermal conductivity of metals with which wood often is mated in construction. It is about two to four times that of common insulating materials. For example, the conductivity of structural softwood lumber at 12% MC is in the range of 0.10 to 0.14 W/(m·K) compared with 216 for aluminum, 45 for steel, 0.9 for concrete, 1 for glass, 0.7 for plaster, and 0.036 for mineral wool.

k of wood is affected by a number of basic factors: specific gravity, MC, extractive content, grain direction, structural irregularities such as checks and knots, microfibril angle, and temperature. k increases as density, MC, temperature, or extractive content of the wood increases. In the radial and tangential directions, k is nearly the same. However, conductivity along the grain has been reported around 1.8

times greater than that across the grain.

Heat capacity

Heat capacity is defined as the amount of energy needed to increase per unit of temperature. The heat capacity of wood depends on the temperature and MC of the wood but is practically independent of density or species. The heat capacity of wood that contains water is greater than that of dry wood. Below FSP, it is the sum of the heat capacity of the dry wood and that of water and an additional adjustment factor that accounts for the additional energy in the wood–water bond.

Another term about the heat capacity is the specific heat, which refers to the heat capacity of a unit mass of the material. It depends on the temperature and MC of the wood, but is practically independent of density or species. When wood contains water, the specific heat increases because the specific heat of water is larger than that of dry wood.

Thermal diffusivity

Thermal diffusivity is a measure of how quickly a material can absorb heat from its surroundings. It is defined as the ratio of thermal conductivity to the product of density and heat capacity. Therefore, conclusions regarding its variation with temperature and density are often based on calculating the effect of these variables on heat capacity and thermal conductivity. Because of the low thermal conductivity and moderate density and heat capacity of wood, the thermal diffusivity of wood is much lower than that of other structural materials, such as metal, brick, and stone. A typical value for wood is 1.6×10^{-7} m^2/s, compared with 1×10^{-5} m^2/s for steel and 1×10^{-6} m^2/s for stone and mineral wool. For this reason, wood does not feel extremely hot or cold to the touch as do some other materials.

Coefficient of thermal expansion

The coefficient of thermal expansion is a measure of the relative change of dimension caused by temperature change. The thermal expansion coefficients of completely dry wood are positive in all directions; that is, wood expands on heating and contracts on cooling. Limited research has been carried out to explore the influence of wood property variability on thermal expansion. The thermal expansion coefficient of oven-dried wood parallel to the grain appears to be independent of specific gravity and species. In tests of both hardwoods and softwoods, the parallel-to-grain values have ranged from about 3.1×10^{-6} to 4.5×10^{-6}/K.

Thermal expansion coefficients across the grain (radial and tangential) are proportional to specific gravity. These coefficients range from about 5 to more than 10 times greater than the parallel-to-grain coefficients and are of more practical interest.

Electrical properties of wood

The electrical properties of wood depend strongly on MC, exhibiting changes that span almost 10 orders of magnitude over the range of possible MCs. It has been noted in Chapter 3.1 that electrical moisture meters could be used to accurately predict the MC of wood because electrical properties undergo large changes with relatively small changes in MC below FSP. Two general types of electrical moisture meters have been used: **resistance** meters and **dielectric** moisture meters, respectively.

In this chapter, the electrical properties of wood are discussed, including the effects of wood MC and other parameters, as well as their application in electrical moisture meters. The electrical resistance of wood measured with direct current (DC) and the dielectric constant measured with **alternating current (AC)** will be described respectively.

Electrical resistance

Dry wood is an excellent electrical insulator, with a resistivity in the order of 10^{17} $\Omega \cdot$cm at room temperature. However, the resistivity r decreases dramatically as MC increases. In Fig.3.5 a, when MC increases to 7%, r decreases to approximately 10^{11} $\Omega \cdot$cm. In this MC range, it is equivalent to a reduction in r of 5 times for each percent increase in wood MC. In the moisture range from 7% to FSP, the logarithm of r is essentially a linear function of the logarithm of MC (Fig.3.5 b). The linear relation among r and MC in logarithmic scale in this moisture range provides the possibility for predicting wood MCs through resistance moisture meters.

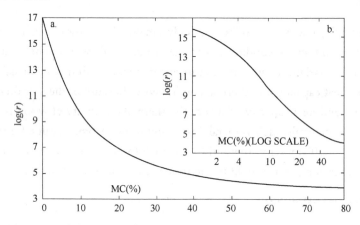

Fig.3.5   Relations of log(r) as a function of wood MC on both linear (a) and logarithmic (b) scale (source: Skaar, 1988)

Increasing the temperature of wood has the similar effect on the value of r as its MC increases. Therefore, when estimating the wood MC by means of a resistance moisture meter, the wood temperature should be obtained previously, as well as the temperature correction factor dMC/dT. This empirical factor gives the apparent increase in wood MC, as read on the moisture meter, per degree increase in wood temperature above the calibration temperature for the meter. The factor dMC/dT is considered to describe the effect of temperature on electrical resistivity in terms of the activation energy E. The activation energy is a useful concept in investigating the mechanism of electrical conduction in wood.

Except for MC and temperature, the anatomical structure of wood influences the electrical resistivity as well. According to reports, the longitudinal conductivity is 1.9 to 3.2 times greater than the radial direction, and 2.1 to 3.9 times greater than the tangential direction. In other words, the electrical resistivity in the transverse direction is significantly smaller than that in the longitudinal direction. The variation of electrical conductivity among three major directions is attributed to the difference of conductivity between substances within the cell wall and the air in the cell lumen. As for transverse directions, the cross walls, portion of the cell wall perpendicular to the direction of the electric field, are relatively ineffective in conducting electrical current compared with the parallel walls.

It has been also indicated that considerable variation of electrical properties among wood species may be attributed to different amounts of chemical constituents of wood, including extractives and mineral constituents. Generally, the wood species with higher lignin contents tend to have higher conductivities at both oven-dried and air-dried states.

In addition, the electrical conductivity is directly proportional to ion content in wood because the charge carriers are believed to be primarily monovalent ions (Na^+, K^+, etc). When the wood is treated with preservative or fire-retardant, or stored in salt water for extending time periods, the electrolyte content may be increased significantly, and resistance moisture meter readings tend to be high. While, some organic preservatives such as creosote have the opposite effect in that they decrease the r-value of wood, thus giving a low estimation of wood MC.

Dielectric resistance

When an AC is applied, the dielectric constant of wood can no longer be represented by a scalar, because the response will be out of phase with the original signal. The AC dielectric constant is a complex number $\varepsilon^* = \varepsilon' + j\varepsilon''$ with real component ε', imaginary component ε'', and imaginary number j. Instead of presenting the real and imaginary components of the dielectric constant, the loss tangent $\tan\delta$ was derived by

$$\tan\delta = \frac{\varepsilon''}{\varepsilon'} \tag{3-4}$$

The loss tangent $\tan\delta$ represents the ratio of dissipated energy to stored energy. Both ε', ε'' and $\tan\delta$ are all frequency – dependent for wood and other dielectric materials when containing polar molecules. Depending on measurement frequency, at least four kinds of polarization occur in wood. They are electronic, atomic, dipole, and interfacial polarizations.

The dielectric properties of wood at a given MC could be regarded as the multi-effects of both water and dry wood. The dielectric constant of liquid water decreases from near 80 below 10^9 Hz to about 5, above 10^{12} Hz at room temperature. The dielectric constant ε' for dry wood is approximately 4 to 5 in the radio frequency range, varied with a number of intrinsic factors, such as density, grain orientation, and chemical composition, as well as testing temperature and frequency. Concerning the dielectric constants of dry wood and of pure water are considerably different, therefore, the moisture has a pronounced effect on the dielectric properties of wood. Moisture meters have been devised by this principle to measure the MC of wood. The effect of water on dielectric properties of moist wood could be divided as two aspects: the first aspect is that the dielectric constant of wood should increase dramatically with increasing MC; in addition, water also has a pronounced effect on anomalous dispersion in wood.

Electrical moisture meters

As noted earlier, there are two principal types of electric moisture meters, the resistance meter which generally measures the DC resistance of wood between two electrodes, and the dielectric moisture meter which measures one or more of the dielectric properties of wood using AC.

Resistance moisture meters are essentially DC megohm-meters, with scales calibrated in terms of wood MC based on one or more sets of calibration data. Adjustments for different kinds of wood are usually given in tables supplied by the manufacturer. The oldest resistance moisture meter, produced as early as 1930, measured wood resistance in terms of the size capacitor, which could be charged to a given voltage in a period of 1 or 2 seconds, using a standard charging voltage in a series resistance-capacitance circuit. The resistance moisture meters have been used for many years to monitor MC in the lumber drying process. Sometimes, they are connected to electrodes embedded into the green lumber before drying, and the measurement of resistance between electrodes as well as wood temperatures are performed during the drying process.

Concerning that both the dielectric constant and loss factor increase with increasing wood MC at a given AC frequency, there are two basic types of dielectric moisture meters, the capacitance type (which essentially measures the dielectric constant) and the power-loss type (which measures the combined effect of dielectric constant and loss factor). The capacitance type of moisture meter responds primarily to the capacitance between the electrodes, which is a function of the electrode configuration and the dielectric constant of the wood. Usually, there is an air gap between the electrodes and the wood surface which reduces the sensitivity of the instrument to MC changes in the wood. The power-loss type of dielectric moisture meter responds to increases in both the dielectric constant and loss factor with increasing wood MC. In other words, it measures the rate of electrical energy dissipation in the wood for given input energy. This depends on the electrode configuration, which determines the electromagnetic coupling between the meter and the wood, as well as on the characteristics of the wood itself.

Despite the many factors in addition to wood MC which affect dielectric moisture meter readings, they are useful instruments for estimating wood MC. They are often used in the wood industry to indicate wood MC during processing operations under reasonably standard conditions of species.

Acoustical properties of wood

Solid wood and some wood-based composites are considered as **acoustic** materials because of their ability to absorb an important amount of incident sound in order to reduce the sound pressure level and the reverberation time in a room. They are applied to walls and ceiling surfaces or to floor platforms, depending on the performance requirements of the room space, for speech and music listening, in offices, industrial buildings, homes, etc. In addition, acoustic properties are also used to evaluating the quality of wood, and even logs. This section would describe the theory and experimental method for the sound absorption characterization of wood, and the influence factors will also be presented.

Fundamentals of wave propagation in wood

When a long and slender wood specimen is subjected to applied stress, a wave **propagates** along the longitudinal direction of the specimen in a certain velocity. The wave propagation of wood is dependent on its density and elastic parameters (modulus of elasticity E and Poisson's ratio μ) of wood.

Experimental studies on the acoustic properties of wood and wood-based composites are extensive and some typical methods, such as free oscillation methods and forced vibration methods or resonance methods, have been developed for measurements in a wide range of frequencies. Based on these methods, the ambient temperature, wood MC, and wood anatomical structure would influence the acoustic propagation to some extents.

Influence of anatomy structure on acoustic properties

In order to relate the anatomical structure to the acoustical behavior of wood, it is necessary to understand some mechanisms that cells response separated during the propagation of vibration. The study of structural features of wood cell walls shows that longitudinal fibers and tracheids are "tubes" of cellulosic crystalline substance embedded in an amorphous matrix (lignin and hemicellulose). In the radial direction, acoustical waves confront another kind of tubular structure in the presence of rays, but in the tangential direction, an acoustical conducting structure is completely absent.

The sonic or **ultrasonic** energy injecting into a fibrous material couples with each fiber in several modes (longitudinal, flexural, and torsional). The physical properties of the cellular wall such as the density, the modulus of elasticity, etc. and the shape and size of the fibers or of other elements affect the transmitted (ultra) sonic field. Each structural element acts independently like an elementary resonator. The spatial distribution of velocities and frequencies that matched the frequency of natural fibers could explain the acoustical behavior of wood illustrated by its overall parameters.

Influence of MC on acoustic properties

As expected, the value of V is the highest when under dry condition compared with moist conditions. The sonic or ultrasonic velocity decreases dramatically with MC up to FSP, and thereafter the variation is very small. At a low MC ($<18\%$), when water is presented in the cell walls as bound water, the wave stress is scattered by the wood cells and by cell boundaries. The side units of —OH or other radicals of the cellulosic material may reorient their position. At a higher MC but under the FSP, the scattering at cell boundaries could be the most important energy loss mechanism. Above the FSP when free water is presented in cellular cavities, the porosity of the material intervenes as a predominant factor in (ultra) sonic scattering.

Words and Phrases

1. **moisture*** *n* [U]: the presence of a liquid, especially water, often in trace amounts. Small amounts of water may be found, for example, in the air (humidity), in foods, and in various commercial products. Moisture also refers to the amount of water vapor present in the air. 通常表示气态水

2. **fiber saturation point (FSP)***: the moisture content at which the cell walls are saturated with water (bound water) and no water is held in the cell cavities by capillary forces. It usually is taken as 25%~30% moisture content, based on weight when ovendry. 纤维饱和点

3. **equilibrium moisture content (EMC)***: a moisture content at which wood neither gains nor loses moisture to the surrounding air. 平衡含水率

4. **relative humidity (RH)***: the amount of water that is present in the air compared to the greatest amount it would be possible for the air to hold at that temperature. 相对湿度

5. **sorption isotherm***: 吸附等温线

6. **sorption hysteresis**: 吸湿滞后

7. **bound water*** 结合水

8. **swell*** *v&n* [C]: become larger or rounder in size, typically as a result of an accumulation of fluid. 湿胀

9. **shrink*** *v&n* [C]: become or make smaller in size or amount. 干缩

10. **orthotropic** *adj*: Having three mutually perpendicular planes of elastic symmetry at each point. 正交异向性

11. **warp*** *v&n* [C]: a twist or curve that has developed in something originally flat or straight. 弯曲, 歪斜, 扭曲

12. **check*** *v*: cause to crack. 细裂, 开裂

13. **split** *v&adj&n* [C]: a narrow break made by or as if by splitting. 分裂, 分开; 分歧; 裂缝

14. **modulus** *n* [C]: A constant factor or ratio 模数, 模量。复数: moduli

modulus of elasticity (MOE)*: 抗弯弹性模量

modulus of rupture (MOR)*: 抗弯强度

15. **tensile strength**: 抗拉强度

16. **hardness** *n* [U]: 硬度

17. **stress** *n* [U]: 应力

18. **stiffness** *n* [U]: The quality of being firm and difficult to bend or move. 刚度

19. **pitch pocket**: 树脂囊

20. **gum** *n* [U]: 树胶

21. **fire retardant**: A substance or treatment that confers the property of slowing or halting the spread of fire. 阻燃剂、阻燃处理

22. **thermal conductivity**: 导热系数

23. **heat capacity**: 热容量

24. **thermal diffusivity**: 热扩散系数, 导温系数

25. **resistance** *n* [U]: 阻力, 阻抗

26. **dielectric** *adj&n* [C]: Having the property of transmitting electric force without conduction; A medium or substance with a dielectric property; an insulator. 非传导性; 电介质, 绝缘体

27. **alternating current(AC)**: 交流电

28. **acoustic** *adj*: Relating to sound or the sense of hearing. 声学的

29. **propagate** *v*: 1) (with reference to motion, light, sound, etc.) transmit or be transmitted in a particular direction or through a medium. 扩散, 传播; 2) (of a plant or animal) reproduce by natural processes. 繁衍, 增殖

30. **ultrasonic** *adj*: sound waves with a frequency above the upper limit of human hearing. 超声的

31. **specific gravity**: as applied to wood, the ratio of the ovendry weight of a sample to the weight of a volume of water equal to the volume of the sample at some specific moisture content, as green, air-dry, or ovendry. 比重

Notes

1. 标准

为了在一定范围内获得最佳秩序, 经协商一致制定并由公认机构批准, 共同使用的和重复使用的一种规范性文件, 按使用范围划分有国际标准(International standard)、国家标准(National standard)、行业标准、地方标准、企业标准等。

常见的国际/国家标准制定机构或协会有: 国际标准化组织(International Organization for Standardization, ISO)、国际电工协会(International Electrotechnical Commission, IEC)、中国标准化协会(China Association for Standardization, CAS)、美国国家标准学会(American National Standards Institute, ANSI)、美国材料试验协会(American Society for Testing and Materials, ASTM)、共同的欧洲标准化组织(CEN/ CENELEC)、加拿大标准协会(Canadian Standards Association, CSA)、日本工业标准调查会(Japanese Industrial Standards, JIS)、日本农林水产省(Japanese Agriculture Standard, JAS)。

为了减少技术性贸易壁垒、适应国际贸易的需要、提高我国产品质量和技术水平, 在制定我国标准时, 经

过分析研究和试验验证,通常将国际标准的内容等同或修改转化为我国标准。我国标准采用国际标准的程度代号为: IDT(等同采用)、MOD(修改采用)和 NEQ(非等效)。

在本书中,重点专业词汇都辅以英文释义,以便读者更准确地理解它们的确切含义,其中部分来源于以下标准:

1) ASTM D9-78, Standard Terminology Relating to Wood
2) ASTM D907-04, Standard Terminology of Adhesives
3) ASTM D1038-83, Standard Terminology Relating to Veneer and Plywood
4) ASTM D1554-01, Standard Terminology Relating to Wood-Based Fiber and Particle Panel Materials

2. 各向同性、各向异性、正交各向异性

根据物体在不同方向上性能的异同性,可以分为各向同性、各向异性和正交各向异性等:

1)各向同性(isotropic): 某一物体在不同的方向所测得的性能数值完全相同,亦称均质性;

2)各向异性(anisotropic): 某一物体的全部或部分化学、物理等性质随着方向的改变而有所变化,在不同的方向上呈现出差异的性质;

3)正交各向异性(orthotropic): 某一物体在 3 个相互垂直的基准轴方向上的性能都是单值的且相互独立。

木材通常被看做是一种正交异向性材料,其 3 个主方向分别为: 轴向(longitudinal, L)、径向(radial, R)和弦向(Transverse, T)。当描述木材的正交异向性时,涉及 9 个独立弹性常数,分别为弹性模量 E_L、E_R、E_T;泊松比 $\mu_{LR}(\mu_{RL})$、$\mu_{LT}(\mu_{TL})$、$\mu_{RT}(\mu_{TR})$以及剪切模量 G_L、G_R、G_T。因此,描述木材强度的指标一般也需指明方向: 顺纹抗拉强度(Tensile strength parallel to grain)、横纹抗拉强度(Tensile strength perpendicular to grain)、顺纹抗压强度(Compressive strength parallel to grain)、横纹抗压强度(Compressive stress perpendicular to grain)、抗弯强度(Modulus of rupture)、冲击韧性(Impact bending)和硬度(Hardness)。

Exercises

1. Why moisture plays a critical role in wood properties and performance?

2. Describe the moisture adsorption and desorption processes and summarize the factors that affect the equilibrium moisture content of the wood.

3. Why free water and bound water have different influence on the physical and mechanical properties of wood?

References

1. Skaar C. Wood-water relations. New York: Springer-Verlag, 1988.

2. Siau J F. Wood: Influence of moisture on physical properties. Department of Wood Science and Forest Products, Virginia Polytechnic Institute and State University, Blacksburg, 1995.

3. Bucur V. Acoustics of wood. New York: Springer-Verlag, 2006.

Wood Products Manufacturing Process

Lumber manufacturing

Compared to other **manufacturing** processes, lumber manufacturing is fairly simple. After logs are cut from the standing trees, they are transported to a **sawmill** by trucks, where they are graded according to size and suitability for different uses. The logs go through an inspection, **debarking** and camera scanning prior to cutting into lumber. Various types of **saws** could be used to cut the lumber into different lengths, widths, and thicknesses. The boards are then stacked for subsequent processes.

Basic sawing patterns

There is no single best sawing method for all logs. To maximize the volume **yields**, the available saws, the log quality and size, the market demands, and the saw operators should be considered for determining the appropriate sawing method. There are four basic sawing patterns: live-sawing, sawing around, **cant**- and quarter-sawing (Fig.4.1). Sawing around and quarter-sawing are only appropriate for large logs (diameter > 500 mm), while quarter-sawing is used rarely for softwoods. In general, cant-sawing provides higher volume yields than live-sawing because in cant-sawing some **taper** parts in the cant can be recovered as short boards, whereas in live-sawing these taper parts are considered as edgings. Furthermore, there is an increased incidence of large spike knots when applying live-sawing, which results in lower recovery of higher grades in softwoods.

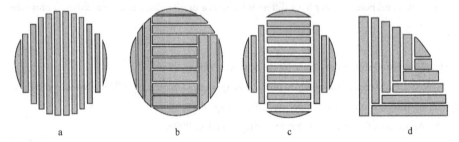

Fig.4.1 Basic cutting patterns

a. live-sawing; b. sawing around or sawing for the grade; c. cant-sawing; d. quarter-sawing

Sawing could involve split-taper (sawing parallel to the central axis of the log) or full-taper (sawing parallel to the cambium). In general, full-taper gives a higher conversion with short logs having little taper, on the other hand, with long logs having significant taper, split-taper sawing gives a reasonable conversion of short side boards and a better conversion from the central cant compared to full-taper sawing. With split-taper sawing, the width of the central cant is approximately the same on both its faces. With full-taper sawing, the cant has one face of constant width along its length, whereas the opposite face is heavily tapered, resulting in a lower conversion.

Basic sawing types

There are several saws used to break logs into boards or larger dimension timber: **circular saw**, **bandsaw**, **frame saw** and **chipper canter**. Sawmills use a variety of saws to progressively cut the logs into the timber of the desired dimensions. The choice of machinery is influenced by log quality, size, and volume.

A circular saw blade rotates on an arbor and is self-supporting. A saw blade must be correctly tensioned to ensure that it cuts true. **Tensioning** involves carefully hammering or squeezing the blade between narrow rolls in a zone about half to two-thirds of the saw blade from the center to the periphery so that the metal there is very slightly thinner and instead is spread out sideways. The greater the degree of tensioning the greater the dishing effect. When the saw is running at the desired peripheral speed, the metal at the rim stretches under the centrifugal force, while the metal nearer the collar can expand correspondingly, counter-balancing the strains induced by hammering. The saw blade straightens and can cut accurately when running at the correct peripheral speed (generally 30 ~ 45 m/s). Small circular saws are used in most mills to edge the boards and timber, but it is undesirable to use circular saws to process large diameter logs or thick timber. A double arbor circular saw allows a greater depth of cut without quite such an excessive kerf. Two saws cut in exactly the same plane, but the upper saw is offset (ahead of the lower saw) so the teeth do not mesh. Any lateral offset between the saws produces a small step on the cut faces which must be planed off later.

A bandsaw has an endless steel band which is mounted between two large wheels. These wheels can be 1.5 to 3 meters in diameter, with larger wheels being used on the **headrig** and smaller wheels being used on re-saws. The lower, heavier wheel is powered and pulls the blade down through the log as it is fed into the saw. The blade is strained between the two wheels and the blade can be of thinner gauge metal than that used in a circular saw. Bandsaws still require tensioning as the teeth get warm as they cut. If being poorly tensioned, the blade will wander, and the timber will have to be dressed afterward to get a good finish. Bandsaws are ideal for making deep cuts with small kerf because of the length of the saw (as much as 10 to 20 m). They can cut accurately and run for long periods between sharpening. Bandsaws can be used as headrig or re-saws.

A framesaw (or gangsaw) can be fitted with several blades at any desired spacing. The thickness of the two slabs, the flitches and the central cant are varied by adjusting the distance between blades within the frame. A continuous feed system is made by oscillating in a figure of eight so that the saw blades move forward through the timber during the downstroke and pull back and away during the upstroke. One drawback with the frame saw is the large inertial forces that limit the speed of the sawblades and so the feed speed. A feed speed of 5 cm/s with large logs or 30 cm/s with small logs is typical. Framesaws require logs of uniform quality and size because it is not possible to turn the log and explore ways of maximizing recovery of the better grades of timber.

Wood drying

After the cutting of green trees, free water and bound water would be removed from cell lumen and cell wall successively by the natural or artificial method. The water removing process is called wood drying, which is one of the most important and effective measures for the protection and preservation of wood.

Basic principles of wood drying

Water in wood normally moves from higher to lower zones of moisture content (MC). The process of drying could be divided into two phases: movement of water from the interior to the surface of wood, and the removal of water from the surface. Water moves through the wood as liquid or vapor through several kinds of passageways. These are the cavities of fibers and vessels, ray cells, pit chambers, and their pit membrane openings. Most water is lost during drying through water removal from cell cavities and pits. Water moves in the passageways in all directions, both along and across the grain. Lighter species are generally dried faster than heavier species because their structure contains more openings per unit volume.

During wood drying, several forces may be acting **simultaneously** to move water:

1. **Capillary** action causes free water to flow through the cell cavities and pits;

2. Differences in relative humidity (RH) cause water vapor to move through the cell cavities by diffusion, which moves water from areas of high RH to areas of low RH. Cell walls are the source of water vapor, that is, water evaporates from the cell walls into the cell cavities;

3. Differences in MC cause bound water to move through the cell walls by **diffusion**, which moves water from the area of high MC to areas of low MC. Generally, water molecules move through both cell walls and cell cavities of wood by diffusion. Water may evaporate from cell walls into cell cavities, move across the cell cavities, be re-adsorbed on the opposite cell walls, move through the cells by diffusion, and so on until it reaches the surface of the board.

When green wood starts to be dried, evaporation of water from the surface cells by capillary forces, pulling free water to flow in the zones beneath the wood surface. Longitudinal diffusion is about 10 to 15 times faster than lateral (radial or tangential) diffusion. Radial diffusion, perpendicular to the growth rings, is somewhat faster than the tangential diffusion, parallel to the rings. This explains why the **flatsawn lumber** dries faster than the **quarter-sawn lumber**. Although longitudinal diffusion is 10 to 15 times faster than lateral diffusion, it is of no practical importance, because in most drying cases, the longitudinal dimension of lumber is much greater than the lateral dimension. During drying, therefore, most of the water is removed through the thickness direction, leaving from the wide face of a board. Since the width and thickness are not greatly different in dimension lumber, such as in squares, significant drying occurs in both the thickness and width directions. The rate of diffusion depends mostly upon the permeability of the cell walls and their thickness. Thus, permeable species dry faster than impermeable ones, and the rate of diffusion decreases as the specific gravity increases.

When water removes from the wood, the MC at the surface of wood reduces at first. The MC gradient occurs along with the thickness of wood, and the shrinkage starts in the outer areas before it starts in the inner areas. Drying stresses develop due to the shrinkage variation in different areas (Fig.4.2). Drying stress is one of the main reasons causing defects, such as cracks and warp. Understanding the development of drying stress is essential for reducing specific wood defects and improving the quality of wood products. Most defects that develop in wood products during and after drying could be classified as the rupture of wood tissue, warp, uneven moisture content, and discoloration. When drying stress at the surface of the board exceeds the tolerant tensile strength of the wood perpendicular to the grain, a **surface check** occurs usually in wood rays. Surface checks are developed because the lumber surface loses water too fast due to the low RH in a dry **kiln**. Thick, wide, flatsawn lumber is more susceptible to surface checking than thin, narrow lumber. Tension drying stress that develops in the core of boards during drying may cause inner checks (or **honeycombing**).

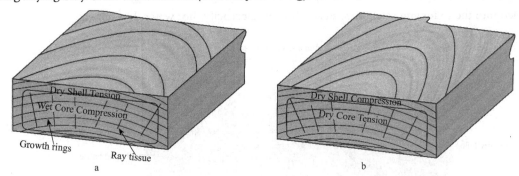

Fig.4.2　End view of a board showing development of drying stresses at early (a) and later (b) period in drying

In addition, there is another cause of drying stress: hydrostatic tension. Hydrostatic tension forces develop during the flow of capillary water. As water evaporates from cell cavities near the wood surface, the water in the wood core is pulled out. This tension pull is inward on the walls of cells whose cavities are full of water, and the pulling result can cause a collapse of the cell wall. The **collapse** is more likely to occur at the early stage of drying when many cell cavities are full of water and the drying temperature is high.

Air drying

As long as liquid water is present in wood, a step moisture gradient exists. Its magnitude could be estimated approximately as the double ratio of "difference of the MC in the wood core and the corresponding equilibrium moisture content (EMC) with the surrounding atmosphere" to "the thickness of the wood". The rate of air-drying is directly proportional to this gradient. The average EMC values in most areas of China varies from 10% to 16%. Important factors influencing the EMC are climate, geographical site as well as elevation above sea level, the prevalence of dry or wet winds, the tendency for fog (e.g. near rivers, lakes or swamps), the amount of precipitation, and average duration of sunshine. All these conditions may vary considerably not only seasonally, but also from year to year and from place to place.

Though air **seasoning** is dependent mainly on the climate and the seasons, the yard is a key factor for the success of air drying. The piles should be arranged in such a way in the lumber yard that an excellent circulation of the air is secured. The piling methods have an effect on the circulation. Normally, the wide alleys should run parallel to the prevailing wind, but if enough space is available the wide alleys should run both along and across, thus facilitating good circulation through the yard.

The main purpose of air-drying in sawmills is a reduction of the MC of the lumber to about 25%~30%. In this condition, the lumber is ready for transport, but not low enough MC for most structural application purses.

Kiln drying

A lumber dry kiln (Fig.4.3) consists of one or more chambers designed to provide and control the environmental conditions of heat, humidity, and air circulation necessary for the proper drying of wood.

The drying temperature range depends largely on the species to be dried and quality and end use of final products. The amount of production expected, source of energy, and limitations of certain components of the system are also considered. According to operating temperature range, the kiln drying is classified as low-temperature drying (<50°C), conventional-temperature drying (50 ~ 80°C), and high-temperature drying (>80°C), respectively. The heating type of kiln drying and the energy source for that heat can be divided into the following categories: steam, hot air, electricity, hot water, hot oil, etc.

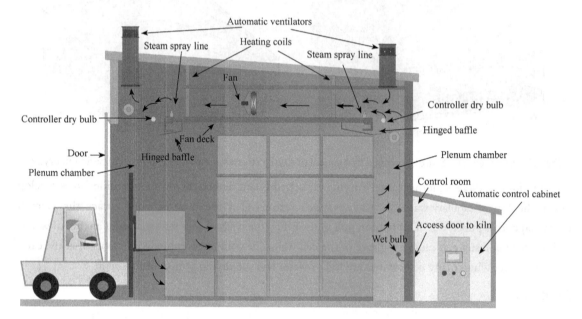

Fig.4.3 A typical package-type dry kiln for hardwood lumber

Kilns are constructed of several materials, including aluminum prefabricated panels, concrete, brick, wood, and plywood. Various kinds of vapor barriers are used to restrict the movement of water vapor from inside the kiln into the structural members and panels and thus prevent deterioration of the structure. To have acceptable efficiency, kilns must be reasonably well insulated against loss of heat through the structure. In addition, doors and other openings must fit tightly to minimize loss of heat and humidity. The choice of building materials is frequently governed by such things as operational temperatures required for the species and thicknesses, life expectancy of the kiln, capital investment, insurance, source of energy, and type of heating system. Except for the heating system, air-circulation system and humidification system should also be concerned. In all modern kilns, adequate circulation is secured by means of fans or blowers. The humidity is controlled by regulating the steam valves, steam sprays, and dampers. When the dampers are open, the humidity goes down; when they are close, the humidity increases. It is possible to regulate the dampers automatically by means of kiln controllers.

The wood drying process should be operated by a **kiln schedule**. The kiln schedule is a carefully worked-out compromise between the need to dry lumber as fast as possible and, at the same time, to avoid severe drying conditions that will cause drying defects. It is a series of **dry- and wet-bulb temperature** sensors that establish the temperature and RH in the kiln and are applied at various stages of the drying process. The schedules must be developed so that the drying stresses do not exceed the strength of the wood at any given temperature and MC. Otherwise, the wood will crack either on the surface or internally or be crushed by forces that collapse the wood cells.

Other drying methods

Since wood drying has been recognized as a vital element in the wood value-added processing, an emphasis has been given during the last several decades to improve wood quality and reduce drying costs. Some novel drying methods, like **solar**, **dehumidification**, **radio frequency**, **vacuum**, and **superheated steam** drying, have been developed and used at a much smaller scale by the wood products industry.

Open-air solar drying has been used since time immemorial to dry plants, seeds, fruits, meat, wood, and other products as a means of preservation. This free and renewable energy source has been used for developing the solar drying technique.

A dehumidification kiln applies the principle of a heat pump. It equips a compressor driving a type of refrigerant in a pipeline first through a cold evaporator coils and then through hot condensing coils. During the drying process, water is removed from the atmosphere by passing warm, moist air from the kiln over the cold **coils**. The energy in the air is captured by the refrigerant, which is then used to heat the dehumidified air before it passes back into the kiln to evaporate more water from the lumber, and the cycle continues. A dehumidification kiln is more energy efficient than a conventional steam kiln because it recycles the heat of evaporation, whereas a steam kiln **vents** this potential energy. The maximum temperature in a dehumidification kiln is limited by the type of refrigerant used and is normally below 70°C which makes it more suitable for hardwood drying.

Radio frequency drying is a method that forming energy instead of forming heat. When wood is placed with a high frequency alternating electromagnetic field, the water dipoles and ions located in the lumens and cell walls oscillate according to the direction of the electromagnetic field, and heating is generated within the wood. When radio frequency drying is combined with a low ambient pressure, the rate of water evaporation from wood could be greatly increased. This method is called as radio frequency/vacuum drying. When low ambient pressure is applied, a pressure gradient, along with the temperature gradient, is developed from the center towards the surfaces of lumber. This pressure gradient helps to drive out wood moisture fast in both liquid and vapor forms. The magnitude of the temperature and pressure gradients could effectively be controlled by the amount of the electromagnetic power, and the ambient pressure inside the radio frequency/vacuum drying kiln.

Superheated steam drying involves the use of superheated steam in place of hot air, combustion, or flue gases as the drying medium to supply heat for drying and to carry off the evaporated water. One of the obvious advantages of superheated steam drying is that the energy efficiency is high due to the re-use of the latent heat of the exhaust. In air drying, the latent heat in the exhausted steam is generally difficult and expensive to recover. In addition, higher drying rate and quality could be obtained by superheated steam drying resulted from high transfer coefficient of heat and mass.

Wood machining

Sanding

The smoothness of all machined wood and composites which will be stained, polished or lacquered depends on the sanding operation. The smoothness of a sanded surface depends mainly on the size, shape, and quality of the particles of sanding grit and on the running speed of sanding belt and feed speed of workpiece. To prevent scratching, mechanical sanders should have an oscillatory movement except in the case of moulding sanders. Therefore, abrasion techniques must be applied. The task of any abrasive is to cut, and the abrasive particles should be compact and strong, with rough surfaces and sharp edges. The raw materials for the abrasive particles are ground in mills, cleaned by washing and dried, separated from iron by means of magnets and classified according to grain sizes. Average and small grain sizes are chosen for the finest sanding work. Experiments prove that a high surface quality also may be obtained with greater grain sizes if the sanding speed is high enough. The efficiency of sanding can be expressed by the loss of weight of sanded material per unit time. The sanding efficiency of the whole process is influenced by structure, strength and wood extractives.

Bending

Bending (or moulding) of wood in forms is a technique that offers the advantages as: there are no losses through wood waste; the shaping is simpler and can be quicker than by means of the usual woodworking machines; and the strength and stiffness of the bent part is higher than the same properties that shaped by sawing or spindle moulding. It is difficult to bend air-dried wood without plasticizing beforehand. The plasticized treatment is by steaming or boiling. For several special purposes after boiling, a combination of bending and compression is applied at the end of the parts to be bent, such as staves for beer barrels. Other special processes: plasticizing by chemical agents, e. g. urea, liquid ammonia or by high-frequency fields. The reason for the difficulties in bending is the low extension of wood in tension failure. The stresses during bending are less dangerous if the wood has a high capacity for changing its shape under stress. Hardwood species are better suited for bending than all softwood species. In softwood species, the abrupt change of the mechanical properties between earlywood and latewood is probably the main cause for failures in bending. In selecting species of timber for bending the following prerequisites should be taken into consideration: availability of material, strength properties, and bending properties.

Words and Phrases

1. **manufacturing** *n* [U]: the process or business of producing goods in factories制造, 制造业
2. **sawmill*** *n* [C]: a factory where trees are cut into flat pieces that can be used as wood锯木厂
3. **debark** *v.* to move bark (= the hard outer covering) from a tree剥皮
4. **saw*** *n* [C]: 锯子
   bandsaw*: 带锯
   circular saw*: 圆锯
   framesaw*: 框锯
5. **yield** *n* [C]: the amount of profits, crops etc that something produces产量, 收益
6. **cant*** *n* [C]: a log that has been slabbed on one or more sides, usually with the intention of resawing at right angles to the widest sawn face. 毛方
7. **taper** *v.* to become gradually narrower towards one end, or to make something become narrower at one end(使)一段逐渐变得细小
8. **chipper canter:** 削片制材联合机

9. **tension** *n & v* [U]: tightness or stiffness in a wire, rope, muscle etc. 绷紧, 拉紧

10. **headrig** *n* [C]: 主锯

11. **simultaneously** *adv*: somethings happen at exactly the same time 同时地, 同步地

12. **capillary** *n* [C]: the force that makes a liquid rise up a narrow tube 毛细(管)作用, 毛细(管)引力

13. **diffusion** *n* [U]: to make heat, light, liquid etc. spread through something, or to spread like this 扩散, 渗透, 弥漫

14. **flatsawn*** *adj*: the grain pattern resulting when lumber has been sawed in a plane approximately perpendicular to the radius of the log. Lumber is considered flatsawn when the annual growth rings make an angle of less than 45 deg with the surface of the piece. 弦切的

15. **quartersawn*** *adj*: grain pattern in which the wide surfaces of the sawn piece extend approximately at right angles to the annual growth rings. Lumber is considered quartersawn when the rings form an angle of 45 to 90 deg with the wide surface of the piece. 径切的

16. **surface check***: a check occurring on the surface of a piece of wood, usually on the tangential face not extending through the piece. 表裂

17. **honeycombing*** *n* [U]: in lumber and other wood products, separation of the fibers in the interior of the piece, usually along with the wood rays. The failures often are not visible on the surfaces, although they can be the extensions of surface and end checks. 内裂

18. **kiln*** *n* [C]: a chamber used for drying and conditioning lumber, veneer, and other wood products in which the temperature and relative humidity of the circulated air can be varied and controlled, often steam heated and vented. 干燥窑

19. **collapse*** *n* [C]: the flattening of single cells or rows of cells during the drying or pressure treatment of wood, characterized by a caved-in or corrugated appearance. 皱缩

20. **seasoning*** *n* [U]: drying; the term often applied to the process of removing moisture from the wood to achieve a moisture content appropriate for the performance expected of the final product. 干燥

21. **kiln schedule***: in kiln drying, the time schedule of predetermined or actual dry-bulb and wet-bulb temperatures used in drying a kiln charge of lumber or other wood products. 干燥基准

22. **dry-bulb temperature***: temperature of the air as indicated by an accurate thermometer, corrected for radiation if significant. 干球温度

23. **wet-bulb temperature***: the equilibrium temperature of a liquid vaporizing into a gas. With water and air, wet-bulb and dry-bulb temperatures give a measure of the relative humidity. 湿球温度

24. **radio frequency**: 高频

25. **dehumidification** *n* [U]: 除湿

26. **solar** *adj*: using the power of the sun's light and heat 利用太阳光(能)的

27. **superheated steam**: 过热蒸汽

28. **vacuum** *n* [C]: a space that is completely empty of all gases, especially one from which all the air has been taken away 真空

29. **coil** *n* [C]: (换热)盘管

30. **vent** *n* [C]: 排气孔

Notes

1. 词根

词根(root)是基本构词的基本词素, 与词缀相对并携带主要词汇信息。能够独立构词的为自由词根(free

root), 必须与其他词素组合构词的是粘附词根(bound root)。在木材科学与技术中, 有不少使用频繁的词根, 如:
- aqu- 水: aqueous solution 水溶液
- de- 去掉, 取消: debark 剥皮, dehumidification 除湿
- flu- 流: fluid 流体
- hydro- 水: hydrothermal 水热的
- hygro- 湿: hygrothermal 湿热的
- juven- 年轻、年少: juvenile wood 幼龄材
- manu- 手; manual 手工做的
- ori- 升起: oriental strand board 定向刨花板
- prim 第一, 最初: primary wall 初生壁
- rupt- 破: modulus of rupture 抗弯强度
- sect- 切割: cross section 横切面
- un-(uni-) 一: unixial 单轴的
- vac- 空: vacuum 真空
- vari- 变: variation 变异

2. 板材的分类与规格

英语中表示木材的词汇很多, 它们的含义有时互有重合, 但也相互区别, 其中 lumber 表示板材, 是一个较为宽泛的称谓。在木材加工和贸易中不同规格的板材可以细分出一些子类别并有专门的称谓, 一般来说可分为:

1) boards: 名义厚度小于 2 英寸(5 cm)的板材, 其中宽度小于 6 英寸(15 cm)称为木条(strips)。

2) dimension: 名义厚度介于 2~4 英寸(5~10 cm), 宽度大于 2 英寸(5 cm)的板材。

3) timbers: 最小尺寸大于 5 英寸(12.5 cm)的板材。

根据板材的刨光与否, 板材可分为:

1) 刨光板: surfaced lumber 也称为 dressed lumber 或 planed lumber。板材的刨光可以是单面、双面、单边、双边, 或上述的任意组合。

2) 毛板: rough lumber, 指锯解后未经刨光的板材。

Exercises

1. Describe the basic procedures for lumber manufacturing.
2. What is the advantage of wood drying?

References

1. Simpson W T. Dry Kiln Operator's Manual. U. S. government printing office, 1991.
2. Walker J C F. Primary Wood Processing. Chapman & Hall, 1993.
3. Mujumdar A S. Handbook of Industrial Drying. 4th Edition. CRC Press, 2014.

Structural Panels

There are limitations on the maximum cross-sectional size and lengths of sawn timber that can be used as a structural component due to the availability of log sizes and the presence of naturally occurring defects in the timber. These defects can be cut out and the timber reconstituted using engineering wood techniques such as finger jointing to create longer lengths of timber of an assured strength grade or laminating to form a **homogeneous** timber section. Combinations of timber or laminated sections with different materials such as wood-based panels or bonding elements are used to produce **engineered wood products** whose maximum size is limited only by manufacturing. In addition to engineered wood products, there are reconstituted board products which comprise smaller wood-based strands and fibers reformed into panel products. These products have structural applications but can also be used in the furniture making, decoration or packaging industries.

Engineered wood products include a range of derivative wood products which are manufactured by binding or fixing the strands, particles, fibers, **veneers** or boards of wood together with adhesives, or other methods of fixation to form composite materials such as **Cross Laminated Timber (CLT)** shown in Fig.5.1. These products are engineered to precisely designed specifications, which are tested to meet national or international standards. Engineered wood products are used in a variety of applications, from home construction to commercial buildings and industrial products. The products can be used for joints and beams that replace steel in many building projects. These engineered wood products are typically manufactured by adhesively laminating together smaller softwood sections, laminates, veneers or strands of wood.

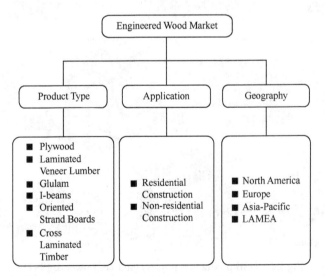

Fig.5.1 The engineered wood products type, application and main markets

Oriented strand board

Oriented strand board (OSB) belongs to the subset of **reconstituted wood panel** products called flakeboard. Oriented strand board originated in the early 1980s. The relatively long and narrow wood or bamboo **strands** are blended with resin and formed into three or five layered mats. OSBs are mainly developed from wood strands, which are typically 15 ~ 25 mm wide, 75 ~ 150 mm long and 0.3 ~ 0.7 mm thick, being cut from logs with small diameters. The adhesive is used to bond the strands together and the boards are fabricated under pressure at a high temperature. The strands in the outside layers are aligned parallel to the board length, whereas the internal strands are deposited perpendicular to the face layers. Aligning the strands in each layer perpendicular to adjacent gives OSB **flexural properties** superior to those of randomly oriented boards. Oriented strand board is produced from either hardwoods or softwoods. Softwoods generally correspond to coniferous species. The most commonly used softwoods for manufacturing OSB are pine, fir, and spruce. Hardwood generally corresponds to **deciduous species**. **Aspen** (*Populus* sp.) is the most commonly used hardwood species for manufacturing OSB.

Fig.5.2 shows the production process of OSB. During the OSB manufacturing, whole logs with small diameters are debarked and cut to 2.5 m length. Then the **waferizer** slices the logs into the appropriate strand sizes. The strands must pass through screens to remove fines and **differentiate** core and surface strands, which being conveyed directly to wet particle storage bins to wait for processing through the dryers.

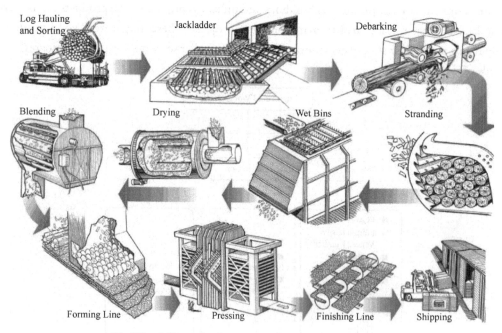

Fig.5.2 OSB production process (source: Lowood, 1997)

Triple pass rotary drum dryers are typical in OSB plants. The rotary dryers are normally fired with wood residues from the plant, but occasionally oil or natural gas is also used as fuels. Conveyor dryers may also be used to dry strands. Regardless of dryer type, strands are dried to a low moisture content of about 6% to 8% to **compensate** for moisture gained by adding resin and other additives. Generally, dryers are dedicated to drying either core or surface material to allow independent adjustment of moisture content, this independent adjustment is particularly important where different resins are used in core and surface materials. After drying, the dried strands conveyed **pneumatically** from the rotary dryer are separated from the gas stream at the **cyclone**. Then the screening is used to remove fines which might absorb too much resin and to separate the strands by different sizes and surface areas. The dried and screened strands are then conveyed to the **blender**, where they are blended with resin, wax and other functional additives. The most commonly used binders are thermosetting **phenol formaldehyde (PF)** and **isocyanate** resins. Then, the glued strands are moved to the former. The strands are mechanically oriented in one direction as they fall through the screen below. Subsequent forming heads form distinct layers in which the strands are oriented perpendicular to those in the previous layer. In the mat trimming section, the continuously formed mats are cut into the desired length by a traveling saw. The mat is then sent to the **multi-opening hot press**. The press applies heat and pressure to **cure** resin and bond strands into a solid panel. Boards exiting the press are cooled and then trimmed the panel to the desired dimension.

OSB is, therefore, a multi-layer board, being used in a large variety of structures of an industrial or decorative style as shown in Fig.5.3. The boards are used for covering floors, ceilings and sometimes walls. The OSBs are also utilized in more and more areas, such as packaging material, wall decorating etc. In recent years, China has developed a surface-finished oriented strand board, which is mainly used in the field of furniture.

Fig.5.3 The surface pattern of OSB

Laminated veneer lumber

Laminated veneer lumber (LVL) is made of veneers with a thickness between 2 and 4 mm bonded together with the same orientation in the layers. The start of the LVL manufacturing process sometimes is consistent with the production of common plywood. The process includes log debarking, cutting to length, steaming and peeling to veneers. The veneer dryers used are the same types of veneer dryers for plywood production. Both hardwood or softwood veneers for LVL are dried at a typical drying temperature of around 180 ℃. The veneer dryer may be a longitudinal dryer or a jet dryer. Once the veneers are dried, they are graded **ultrasonically** for stiffness and strength. The lower grade veneers are used for the LVL core and the higher grade veneers are used in the LVL surface. Then the veneers are passed under a **roll resin coater or** dispensing gluing system where the phenol formaldehyde (PF) resin or isocyanate resin is commonly applied. Once resinated, the veneers are laid up into a long thick stack. The veneer stack is fed to a hot press where the veneers are pressed into a solid state under heat and pressure. The LVL can be produced either a fixed size using a batch press or to an indefinite length using a **continuous press**. Electricity, microwaves, hot oil, steam, and radio frequency radiation can be used to heat the veneer stack. Fig.5.4 shows the side view of the LVL samples.

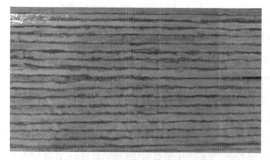

Fig.5.4 LVL longitudinal side

The different stages in the manufacture of LVL are intended to eliminate the defects **inherent** in ordinary sawn timber and to develop higher strength, better dimensional stability and more homogeneous physical and mechanical properties along the beam or elements. LVL is in most cases bonded with the PF or **melamine- formaldehyde (MF)** adhesive and has higher strength and stiffness than plywood. LVL can be found in a large variety of dimensions: length up to 25 m, thicknesses below 150 mm and width up to 2.5 m. LVLs are mostly used as wall studs or floor grilles in the light wood frame construction. LVLs can be used as the wings of the wood I-beam. In China, LVL is mainly used for frame structures of wooden doors and windows owing to its excellent mechanical properties.

Glued laminated lumber

Glued Laminated Lumber (Glulam) is a structural element made of sawn timber glued together with parallel fiber orientation to straight beams or members with some kinds of **curvatures**. Glulam is used in a wide variety of applications, ranging from supporting beams in residential framing to major structural elements in non-residential buildings as girders, columns, and truss members. Softwood and hardwood species are being used in glulam, but the most common species for glulam production are **Norway spruce** (*Picea abies*), Douglas fir (*Pseudotsuga menziesii*), larch (*Larix* sp.), **Scots pine** (*Pinus sylvestris*), **southern pines, radiate pine** (*Pinus radiata*) and **yellow poplar** (*Liriodendron tulipifera*).

The major advantage of glulam is its high strength and stiffness, which makes it possible to manufacture structures with wide spans and enhanced bearing capacity. The glulam manufacturing process mainly consists of four phases: 1) drying and grading the lumber; 2) end finger jointing the lumber into longer laminations; 3) gluing the laminations and 4) fabrication and finishing.

Lumber used to produce glulam may be dried in lumber dry kilns or purchased pre-dried lumber form from suppliers, the moisture content of the lumber entering the glulam manufacturing process should be checked. Once the lumber has been examined for moisture content, knots appearing on the lumber should be **trimmed off** and the lumber is graded. The lumber is sorted into stacks based on its grade. The main production processes of glulam are shown in Fig.5.5. The cross-section of the glulam can be built up from lamellae with approximately the same strength, so-called homogeneous glulam. In order to utilize the timber's strength in the best way, high-strength sawn timber is used for the outer parts of the beam where the stresses are highest, so-called combined glulam.

To manufacture glulam in lengths beyond those commonly available for sawn lumber, the lumber must be end-jointed. The most common end joint is finger joint. The finger joints are machined on both ends of the lumber and then structural resin is applied and cured under end pressure. Most manufacturers use a continuous **radio frequency curing system** to cure end joints. Of course, the longitudinal length of the wood can also be mitered. However, it is necessary to ensure a sufficient ratio of the length of the bevel to the thickness of the jointed lumber in order to obtain better mechanical performance.

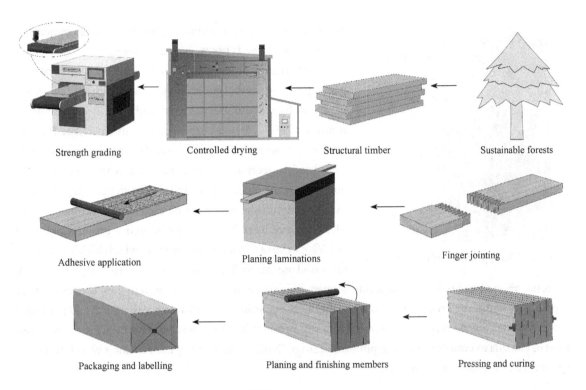

Fig.5.5 Glulam production process

After that, the end-jointed lumber is planned on both sides to ensure a clean surface for gluing. The resin can be spread onto the lumber with a glue extruder. The resinated lumber is assembled into a specified lay-up pattern. Common straight beams are clamped in a clamping bed where a mechanical or **hydraulic system** brings the lumber into close contact. Cured beams are clamped in a curved form. Glulam can be cured at room temperature for 5 to 16 hours before pressure is released. After the glulam beams are removed from the clamping system, the faces are planned or sanded to remove beads of resin that have squeezed out between the boards. Lastly, packaging and labeling should be completed. In glulam manufacture, only **adhesives** that have high strength and durability under long-term loads are used. At the same time, it is necessary to ensure that the uniform distribution of the long jointing points and the lumber combination meet the relative standard. Otherwise, the overall performance of the glulam is greatly reduced and it is very dangerous to apply in the construction environment. Glulam is in line with the development trend of China's fabricated buildings and will produce a great market impact in China.

Parallel strand lumber

Parallel strand lumber (PSL, also called Parallam) is manufactured by gluing strips of veneer together with the grain of each veneer piece oriented to the length direction of the finished product shown in Fig.5.6. Both hardwood such as yellow poplar and softwood such as Douglas fir, western hemlock, and southern pines are used to produce PSL. The manufacture of PSL is based on the technology which makes it possible to convert small trees into elements with large cross-sections and considerable lengths. These

Fig.5.6　The appearance of the PSL pillar

products are intended for building construction as elements in compression, large **trusses**, beams or posts.

The manufacturing process begins with the **rotary peeling** of logs into veneer about 3 mm thick. The green veneer is clipped into sheets, sorted, and dried in a veneer dryer at around 200 ℃. Veneer dryers used to manufacture PSL are similar to veneer dryers used to manufacture plywood or LVL. The dried veneers are clipped into strands approximately 20 mm wide. One advantage of PSL is that the pieces of veneers smaller than full-size sheets can be used for PSL production. So Many scrap veneers that could not be used to produce plywood or LVL can be used for PSL production which highly improves the wood use ration. The veneer strands are mainly coated with PF resin, aligned and fed into a continuous press. Sometimes microwave is used to cure the resin. After the **hot pressing**, the billets are cut into smaller sizes according to customer specification and packaged for transportation. At this stage, the output of this product in China is relatively small. But it has been used in Canada to contribute a lot of public buildings. With the development of China's wood structure, the future is also promising in China.

Cross laminated timber

Although being as a relatively new product, cross laminated timber (CLT) has its origins in the traditional timber technologies of central Europe and Scandinavia. The evolution of the concept of a composite with rigidly bonded crosswise layers from research to a fully realized construction product happened during the 1990s. Early engineering research occurred firstly in Switzerland and then in Austria. In the early 2000s, manufacturing and construction techniques had matured enough for full-scale production to begin. A significant quantity of CLT buildings has been constructed during the past decade including schools, galleries, public buildings, and houses. Among them are a couple of buildings grabbed media attention such as Kingsway School, Stadthaus Murray Grove, and Bridport House.

CLT is a straight or curved, multi-layer timber member consisting of at least three layers, of which at least one of three are **orthogonally** bonded, which always include timber layers and may also include wood-based panel layers shown in Fig.5.7. The main species currently used for CLT is spruce. Though Scots pine, larch, and Douglas fir are also available.

Fig.5.7　Cross section of CLT

The production process of CLT is in most steps largely comparable with the one of glulam. Basically the production of CLT can be divided into the following steps: 1) strength or stiffness grading of already dried boards, 2) cutting out of local growth defects which do not meet the requirements of the strength class and finger jointing of the board segments, 3) division and cutting of lamellas for later use in longitudinal and transverse layers of CLT, adhesive bonding of lamellas to single layer panels, 4) assembling and adhesive bonding of lamellas or single layer panels to CLT, and 5) cutting and joining to structural elements. CLT is a material for the manufacture of structural elements for use in buildings and bridges.

Wood structure adhesives

Phenol formaldehyde (PF)

PF polymers are the oldest class of synthetic adhesives and was developed at the beginning of the twentieth century. These adhesives are widely used in both laminations and composites because of their outstanding durability, which derives from their good adhesion to wood, the high strength of the polymer and the excellent stability of the adhesive.

PF adhesives are typically used in the manufacture of construction plywood and OSB where exposure to weather during construction is a concern. The PF adhesives can serve for almost all wood-bonding applications, as long as the adhesive in the assembly can be heated. In many cases, if moisture resistance is not needed, a lower cost UF or MUF adhesive can be used.

Isocyanate

Several classes of adhesives used in wood bonding involve the use of isocyanates, because of their reactivity with groups that contain reactive hydrogens, such as **amine and alcohol groups**, at room temperature. This allows great flexibility in the types of products produced because they can self-polymerize or react with many other monomers.

Polymeric diphenylmethane diisocyanates (pMDI) are commonly used in wood bonding and are a mixture of the monomeric diphenylmethane diisocyanate and methylene-bridged polyaromatic polyisocyanates. The higher cost of the adhesive is offset by its fast reaction rate, its efficiency of use and its ability to adhere to difficult-to-bond surfaces. pMDI adhesives are used as an alternative to PF adhesives, primarily in composite products fabricated from strands, and they are sometimes used in core layers of strand-based composites, with a slower curing PF adhesive in the surface layers. The use of pMDI requires special precautionary protective measures because the uncured adhesive can result in chemical sensitization of people exposed to it. A cured pMDI adhesive poses no recognized health concerns.

Epoxy adhesives

Epoxy adhesives are two-component thermoset polymers, including an epoxy, epoxide or ethoxyline group resin and a hardener. The advantages of epoxy adhesives are their good chemical and thermal resistance and low clamping pressure. The curing agent, the hardener, produces an insoluble, intractable, cross-linked thermoset polymer. The properties of the cured epoxy adhesive depend on the type of hardener and the cure temperature.

Epoxy adhesives are currently mostly used for the fiberglass reinforcement of, e. g. , wooden boats and glulam. Their advantages are stronger bonds with the wood, higher durability and greater impact resistance than polyester adhesives, and relatively easy to work with. As for disadvantages, epoxy adhesives are rather expensive and have long curing cycles. Also, most of the epoxy suffers from "amine blush", i.e., after application and during the curing process the epoxy releases a blush to the surface.

Resorcinol adhesive

Cold-setting adhesives with good water resistance are the resorcinol adhesives. Resorcinol-formaldehyde (RF) and phenol-resorcinol-formaldehyde (PRF) adhesives are mainly used in the manufacture of structural exterior grade joints. Among their net advantages of strong joints when setting at ambient temperatures, RF and PRF adhesives are rather unavailable and are thus expensive.

Words and Phrases

1. **homogeneous** *adj*. 均匀的
2. **engineered wood products**: 工程木制品
3. **veneer*** *n* [C]: thin sheets of wood from which plywood is made; also referred to as plies in the glued panel. 单板
4. **Cross Laminated Timber (CLT)***: 胶合层积木
5. **Oriented strand board (OSB)***: 单板层积材
6. **reconstitute wood panel**: 重组木质板材
7. **strand**: a wood flake having a minimum predetermined length-to-width ratio of 2:1. 长条刨花
8. **flexural properties**: 弯曲性能
9. **deciduous species**: 阔叶树种
10. **waferizer** *n* [C]: 大片刨花削片机
11. **differentiate** *v*. 区分
12. **triple pass rotary drum dryer**: 三通道式干燥机
13. **compensate** *v*. 补偿
14. **pneumatically** *adv*. 空气作用的
15. **cyclone** *n* [C]: 旋风分离器
16. **blender** *n* [C]: 施胶机
17. **phenol formaldehyde (PF)***: 酚醛树脂
18. **isocyanate*** *n* [C]: 异氰酸酯
19. **press*** *n* [C]: an apparatus for applying and maintaining pressure on an assembly of veneers and adhesive in the manufacture of plywood. It may be operated mechanically or hydraulically and the platens may be cold or heated depending on the type of adhesive used. 压机
20. **multi-opening hot press***: 多层热压机
21. **cure*** *v*. to change the physical properties of an adhesive by chemical reaction, which may be condensation, polymerization, or vulcanization; usually accomplished by the action of heat and catalyst, alone or in combination, with or without pressure. 固化
22. **Laminated veneer lumber (LVL)***: 单板层积材
23. **ultrasonically** *adv*. 超声地

24. **roll coater**: 滚筒涂胶机
25. **continuous press**★: 连续热压机
26. **inherent** *adj*: 内在的
27. **melamine-formaldehyde (MF)**★: 三聚氰胺树脂
28. **Glued Laminated Lumber (Glulam)**★: 胶合层积木
29. **curvature** *n* [U]: the state of being curved, or the degree to which something is curved. 弯曲(度)
30. **Norway spruce** (*Picea abies*)★: 欧洲云杉
31. **Scots pine** (*Pinus sylvestris*)★: 欧洲赤松
32. **southern pine**★: 南方松, 美国南方几种松木的统称
33. **radiate pine** (*Pinus radiata*)★: 辐射松
34. **yellow poplar** (*Liriodendron tulipifera*)★: 北美鹅掌楸, 马褂木
35. **trimmed off**: 去除
36. **radio frequency curing system**: 高频固化系统
37. **hydraulic system**: 液压系统
38. **adhesive**★ *n* [C]: a substance capable of holding materials together by surface attachment. 胶黏剂
39. **truss** *n* [C]: 桁架
40. **rotary peeling**: 旋切
41. **hot pressing**★: process for increasing the density of a wet-felted or air-felted mat of fibers or particles by pressing the dried, damp, or wet mat between platens of hot-press to compact and set the structure by simultaneous application of heat and pressure. 热压
42. **orthogonally** *adv*: 正交地
43. **amine and alcohol groups**: 胺类和醇类基团

Notes

1. 人造板基本构成单元的表达

人造板类型多样, 基本构成单元也各不相同, 有些容易区分, 如制备胶合板的单板(veneer)和制备纤维板的纤维(fiber), 而有些则容易混淆, 如同为"刨花板", 普通刨花板中刨花用particle表示, 而定向刨花板中刨花则以strand表示, 它们与木片(flake, chip)之间的界线有时十分模糊。美国材料与试验协会(American Society for Testing Materials, ASTM)以术语标准的形式对不同人造板构成单元进行了明确的定义和清楚的区分:

1) flake(薄平刨花): a small wood particle of predetermined thickness specifically produced as a primary function of specialized equipment of various types, with the cutting action across the direction of the grain (either radially, tangentially, or at an angle between), the action being such as to produce a particle of uniform thickness, essentially plane of the flakes, in over-all character resembling a small piece of veneer.

2) strand(长条刨花): a wood flake having a minimum predetermined length-to-width ratio of 2:1.

3) wood flour(木粉): very fine wood particles generated from wood reduced by a ball or similar mill until it resembles wheat flour in appearance, and of such a size that the particle usually will pass through a 40-mesh screen.

4) wood wool(excelsior, 木丝): long, curly, slender strands of wood used as an aggregate component for some particleboards.

5) wafer(宽平刨花): a wood flake having a predetermined length of at least 30 mm.

6) chips(木片): small pieces of wood chopped off a block by ax-like cuts as in a chipper of the paper industry, or produced by mechanical hogs, hammermills, etc.

7) sawdust(锯屑): wood particles resulting from the cutting and breaking action of saw teeth.

8) shaving [(刨切产生的)刨花]: a small wood particle of indefinite dimensions developed incidentally to certain woodworking operations involving rotary cutter heads usually turning in the direction of the grain; and because of this cutting action, producing a thin chip of varying thickness, usually feathered along at least one edge and thick at another and usually curled.

2. 有机化学物质构词法

很多有机化合物由于结构复杂,因而需要专门的构词法进行命名,目前通行的化学物质构词法是由国际理论与应用化学联合会(International Union of Pure and Applied Chemistry, IUPAC)制定的,它的基本原则是以分子中以单键结合的最长含碳分子链为命名基准,分子中其他通过非饱和键相联的碳或非碳氢组分都以前辍或后辍附加上去。

烷烃是有机化合物的命名基础,在英文中它以-ane后辍来表示,前面加上表示碳原子数量的前辍:

碳原子数	前辍	烷烃名
1	meth-	methane (甲烷)
2	eth-	ethane(乙烷)
3	prop-	propane(丙烷)
4	but-	butane(丁烷)
5	pent-	pentane(戊烷)
6	hex-	hexane(己烷)
7	hept-	heptane(庚烷)
8	oct-	octane(辛烷)
9	non-	nonane(壬烷)
10	dec-	decane(癸烷)

当烷烃分子失去氢原子后形成的烷基以—yl替代-ane表示,如甲基可表示为methyl。分子中其他官能团可由专门的前辍或后辍进行表示,常见的官能团及表达方式有

类别	结构	前辍	后辍
羧酸	—COOH	carboxy-	-oic acid
醛	—CHO	oxo-	-al
酮	—CO—	oxo-	-one
醇	—OH	hydroxyl-	-ol
胺	—N	amino-	-amine
烯	—C=C—	—	-ene
炔	—C≡C—	—	-yne
卤素	F—	fluoro-	—
	Cl—	chloro-	—
	Br—	bromo-	—
	I—	iodo-	—
乙烯基	—CH=CH$_2$	vinyl-	—
苯基		phenyl-	—

当某个基团同时有几个存在于分子链上时, 可用前辍表示其数量:

数量	前辍
2	di-
3	tri-
4	tetra-

根据上述原则可以对英文文献中的有机化学物质进行翻译或对其结构进行识别, 如本章中的 diphenylmethane, 即由 3 个部分构成: di-, phenyl-, methane, 其中 di-表示 2 个, phenyl-表示苯基, methane 表示甲烷, 结合起来该词表示双苯基甲烷。再如木素的基本构成单元苯丙烷, 由上述构词法则表达为 phenyl-propane。

以上只是有机化学词汇构词法的一个简要介绍, 如需进一步了解可参考 IUPAC 的详细说明: https://www.acdlabs.com/iupac/nomenclature/

Exercises

1. What is the engineered wood products and why have engineered wood products rapidly developed in recent years?

2. Make a description of OSB production process.

3. What are the differences between LVL and Plywood?

4. What are the main production processes of Glulam and CLT?

5. Why the strength or stiffness grading is so important for producing most of the engineered wood products?

6. What is your opinion about the engineered wood products development in China or can you give some advice to accelerate the development.

References

1. Sandberg D. Additives in Wood Products—Today and Future Development Environmental Impacts of Traditional and Innovative Forest-based Bioproducts. Springer Singapore, 2016.

2. Youngquist J A. Wood-based composites and panel products. In: Wood Handbook, U. S. Department of Agriculture, Government Printing Office, Madison, 1999.

3. Barbu M. C, Reh R, Irle M. Wood-based Composites Research Developments in Wood Engineering and Technology. IGI Global, 2014: 1-45.

4. Low K, Burns K. Engineered wood products. Agricultural Commodities, 2013, 3(1): 130.

5. Guss L M. Engineered wood products: the future is bright. Forest Products Journal, 1995, 45(7, 8): 17.

6. Adair C. Regional Production & Market Outlook: Structural Panels & Engineered Wood Products: 2001-2006. APA—the Engineered Wood Association, 2001.

7. 徐信武. 人造板工艺学专业名词. 北京: 中国林业出版社, 2013.

Nonstructural Panels

Wood-based panels are made of wood or other plant fibers, processed through a special process, bonded with or without adhesive, and pressed under certain conditions. Today, the wood-based panel industry has become an important branch of the wood industry. The word first rotary cutting machine was invented in 1818, and mass production of plywood began in the late 19th century. In recent years, China's plywood production has consistently ranked first in the world. China mainly uses fast-growing poplar or eucalyptus plywood as core boards and imported veneers as plywood surfaces.

Germany first started construction of particleboard in 1941 and invented the continuous extruder technology in 1948. In the 1950s, the single-layer hot press was produced. On the basis of the British Bartev continuous heating press, a modern continuous press with simple structure and advanced technology was invented, which was widely used in the production of particleboard and medium density fiberboard. Since then, due to the increase in the production of synthetic resins and the reduction in cost, the development of the particleboard industry has been further promoted. China's particleboard production began in the early days of the 1950s. In the past three years, China's particleboard has shown rapid growth owing to the explosive development of custom furniture.

Fibreboard manufacturing begins with paperboard production technology in the paper industry. The first commercialization was soft fiberboard. In 1931, the invention of continuous hot mill promotes the development of wet fiberboard production technology. The United States began producing dry hard fiberboard in 1952. China began to produce wet fiberboard in 1958 and began to develop dry medium density fiberboard in the 1980s. Due to the advantages of the dry process in terms of wastewater and cost, it has been rapidly developed in China. Since the beginning of the 21st century, the output of China's medium density fiberboard has consistently ranked among the highest in the world.

Wood plastic composites (WPCs) are roughly 50∶50 mixtures of thermoplastic polymers and small wood particles. The wood and thermoplastics are usually compounded above the melting temperature of the thermoplastic polymers and then further processed to make various WPC products. It is a relatively new material that has many potential uses. The markets at home and abroad for WPC decking lumber have been expanding recently and new application such as doors and windows are being pursued and studied. Deck handrails and fencing are other markets for WPC industry.

Decorative plywood

Decorative plywood is a rigid board usually composed of an odd number of veneers glued together and the **fiber orientation** of the veneer is perpendicular to the fiber orientation of the adjacent veneers. Plywood mills produce veneers for their own production or purchase from suppliers and, after veneers are peeled, the veneer sheets are cut into the required dimensions. The sheets are dried and glued together in a hot press under high pressure to obtain plywood. The plywood is then trimmed, polished and conditioned to the desired moisture content. Plywood can be sorted into several types of applications, i. e. , constructional (exterior plywood) purposes and interior use (decorative plywood), as well as special applications such as **concrete shuttering**, package plywood, plywood with special surface layers, etc. Board types where veneer sheets are glued to a timber-strip core, i. e. , a core ply-wood or face-glued blackboards, are also assigned to the group of plywood.

Plywood is glued with a thermosetting adhesive. Phenol-formaldehyde (PF) based adhesives are used for exterior-type plywood and **urea-formaldehyde** (UF) reinforced urea adhesives and sometimes natural polyphenols (tannins) mixed with synthetic adhesives are used for interior-type plywood. The spread (amount of adhesive per-unit area of bond-line surface) varies from about 100 to 400 g/m^2 depending on several factors pertaining to the wood species, adhesive and manner of application. The production process of plywood is shown in Fig.6.1.

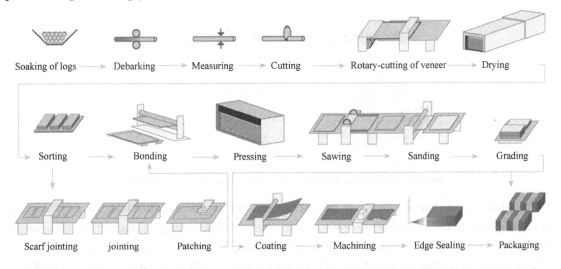

Fig.6.1 Plywood production process (source: WISA Plywood)

Plywood panels may be surfaced with metals, plastics, or other materials, or their veneers may be impregnated to achieve a superficial hardness, or resistance to micro-organisms, fire or other destructive agents. China's plywood production has consistently ranked first in the world for many years, but the overall level of mechanical automation for plywood production needs to be improved.

Particleboard

Generally, **three-layered or graded-structure particleboard** (PB) is manufactured by wood particles that have various dimensions, being cut mechanically by chippers, flakes, etc. The mechanical properties of the boards depend on the particle dimensions as well as on the orientation and arrangement of the wood particles used in the board manufacturing. Of course, the production process also has a great impact on the performance of the final particleboard.

The particleboard industry has been commercialized successfully throughout the world because of the favorite performances of the material as well as the market demands. The PB process makes it possible to use wood trunks of small diameters and **wood residues**. Historically, the availability of raw materials was good and the cost was relatively low. However, this situation has drastically changed in recent years as a result of the rapid growth of the economy and increased market demands. Fig.6.2 shows the main production processes of the particleboard.

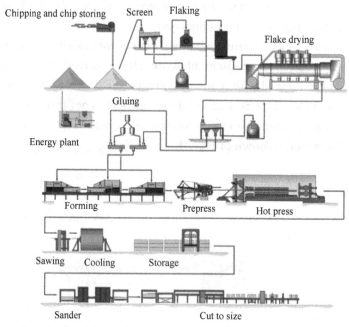

Fig.6.2 Particleboard production process (source: Dieffenbacher)

Particleboards are very useful in the **furnishing and construction industries**. The production of particleboards also leads to effective use of wood with a very small percentage of waste, 5% ~ 15%, instead of 50% in the sawing of logs. The manufacture of particleboard is a dry process and there are two different methods of production: **flat hot pressing and extrusion**, which give different types of boards with different particle orientations. In the first method, the particles are oriented parallel to the panel surface, whereas in the second method the particles are oriented perpendicular to the surface. Fig.6.3 and Fig.6.4 show common hollow core extruded particleboard and its main production mechanism.

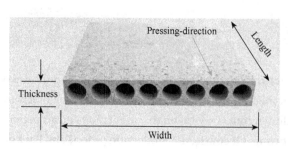

Fig.6.3 Extruded tubular particleboard
(source: Sauerland Spanplatte)

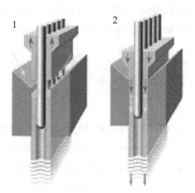

Fig.6.4 Extruded particleboard production
(source: Sauerland Spanplatte)

Flat-pressed particleboard has developed rapidly in China in recent years due to its hot market performance. The particleboard used in custom cabinets is usually covered with melamine impregnated paper, so that eliminates the cost of paint with a beauty and durability performance. Extruded particleboard is mainly used as a core of the sandwich panel for wooden doors. Production in China of Extruded particleboard has decreased due to defects in its mechanical properties and dimensional stability.

Medium-density fiberboard

The essential difference between hard fiberboards and **medium-density fiberboard (MDF)** is that adhesives are used as a binder in MDF, whereas in hard fiberboards the lignin, under the effect of pressure and temperature, is transformed into a kind of adhesive. The MDF manufacturing process began as a **wet or semi-dry process** before being developed into a fully dry process method. The wet production process of fiberboard mainly originates from the production process of papermaking. Due to the problem of sewage generated by the wet production process and the monotonic nature of the product, this process needs to be further improved. A diagram of the dry manufacture of MDF is given in Fig.6.5.

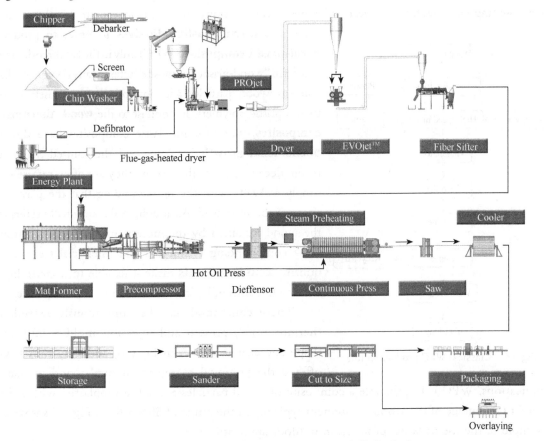

Fig.6.5 MDF production process (source: Dieffenbacher)

Since less water is used than in the wet process, smaller amounts of polluted water are produced.

In addition, this method allows the fabrication of panels with thicknesses from 2 up to 100 mm. A uniform distribution of fibers during manufacture ensures that the MDF has a **homogeneous structure**, and it is possible to manufacture MDF boards with different characteristics to suit particular applications. The MDF is a homogeneous product with a density from 600 to 800 kg/m^3. An MDF board is easy to be machined and its regular surface is exceptionally well suited to painting or the application of a decorative coating. It is this quality which has given MDF the place that it occupies in the furniture industry. The thick MDF is used in joinery and for door frames, window frames, etc. In China, the production of fiberboard and particleboard is much more automated than plywood. Less labor is required for the production of fiberboard and particleboard. Multi-layer hot presses and continuous flat presses are commonly used in Chinese production lines. The newly built production line is mainly continuous flat press owing to a better production process and higher production capacity. China has also developed the 1 mm fiberboard continuous production facility at the world's leading level which can further broaden the use of fiberboard and increase production efficiency.

Wood-plastic composites

Although wood-based composites based on strands, strips, chips, and particles have been made with thermosetting adhesives for many years, it has been made a serious attempt to incorporate wood flour and chips into **thermoplastic adhesives** in order to produce **wood-plastic composites (WPC)** only in the last three decades. The term "wood-plastic composite" refers to any composite that contains wood particles and any of the thermosetting or thermoplastic polymers. In contrast to the **wood- thermoset composites**, wood-thermoplastic composites have shown **phenomenal growth** in China and the United States in recent decades and for this reason, they are often referred to simply as WPC with the understanding that the plastic is always a thermoplastic. Although, WPC originally refers to the wood modified by thermosetting resin impregnation. New compounding techniques and interfacial treatments utilizing **coupling agents** make it feasible to disperse high volume fractions of hydrophilic wood in various plastics.

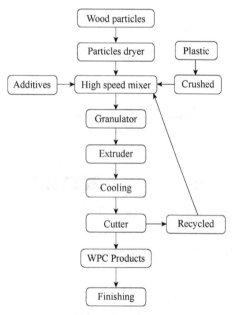

Fig.6.6 Classical WPC production line

These compounds can be continuously extruded, thermoformed, pressed and injection molded into any shape and size as shown in Fig.6.6, and they thus have offered the potential to replace natural wood in many applications. WPC is in principle a composite of wood particles and a thermoplastic, with a dry weight percentage of the wood component typically in the range of 50% ~ 60%. Fig.6.7 shows an example of the use of WPC products in outdoor applications.

Today, WPCs are characterized as a building material and have their main markets in the U. S. and China. The European WPC market is also steadily increasing. In general, WPC products are marketed as a low maintenance building material with a high ability to resist **fungal decay**, although the combinations of wood and polymers often have poor long-term durability when exposed outdoors. A major cause can be the insufficient wood–polymer adhesion due especially to intrinsically low compatibility between the wood substance and the polymers used. Adhesion losses are usually caused by the **hygroscopicity** of wood and the differences in **hygrothermal** properties between the components. At the same time, scientific researchers are constantly developing WPCs with better durability and better mechanical properties.

Fig.6.7 Outdoor application of WPC products

Wood non-structure adhesives

Urea formaldehyde

Urea-formaldehyde (UF) adhesives have several strong positive aspects: very low cost, non-flammable, very rapid cure and light color. UF adhesives are the largest class of **amino resins** and are the predominate adhesives for interior products.

A major drawback of UF adhesives is their poor water resistance and high bond-line failure under **accelerated ageing tests**. Another concern is the long-term **hydrolytic instability** of these adhesive polymers. UF adhesives are believed to gradually depolymerize during their service, resulting in the **continuing emission of formaldehyde.** UF adhesives are typically used in the manufacturing of products used in interior applications, primarily plywood, particleboard, and MDF, because moisture exposure leads to a breakdown of the bond-forming reactions. Excessive heat exposure will also result in the chemical breakdown of cured UF adhesives.

Melamine formaldehyde

Like formaldehyde adhesives made with phenol, Melamine–Formaldehyde (MF) adhesives have high water resistance and are much lighter in color than other adhesives. MF adhesives are most commonly used for exterior and semi-exterior plywood and particleboard, for finger joints, for **decorative laminates**, paper treating, and paper coating. MF resins are often used in combination with UF.

The limit application of the MF adhesives is because of their high cost due to the cost of the melamine. The MUF adhesives, depending on the melamine-to-urea ratio, can be considered as a less expensive MF adhesive. Although the water resistance of MUF is poorer than MF, the substantially lower cost makes it easily accepted by the composite industry. The MUF adhesives can replace other adhesives for some exterior applications.

Soy protein adhesives

As the most important constituent in all living species, proteins are linear **polyamides,** which are built up by amino acids, linked together with polypeptide bonds and together with DNA, fat, and polysaccharides. Protein adhesives allowed the development of bonded wood products such as plywood in the early 20ᵗʰ century. **Petrochemical-based adhesives** replaced proteins in most wood bonding applications because of the lower cost, improved production efficiency, and enhanced durability. However, environmental concerns have led to the reuse of proteins, especially soy flour, which can be used as an important adhesive for interior nonstructural wood products. Generally, soy protein adhesives suffer from high viscosity, consequently demanding low solid content and commonly only meeting the requirements for indoor applications due to its poor water resistance. Extensive research has been conducted to improve the bonding performance and water resistance of proteins to extend the applicability of wood bonded protein adhesives.

Words and Phrases

1. **decorative plywood**: 装饰胶合板
2. **fiber orientation**: 这里特指木材纤维的方向
3. **concrete shuttering**: 混凝土模板
4. **urea-formaldehyde (UF)***: aqueous colloidal dispersion of urea-formaldehyde polymer which may contain modifiers and secondary binders to provide specific adhesive properties. 脲醛树脂
5. **particleboard*** *n* [C]: a generic term for a composite panel primarily composed of cellulosic materials, generally in the form of discrete pieces or particles, as distinguished from fibers, bonded together with a bonding system, and that may contain additives. 刨花板
6. **three-layered or graded-structure particleboard**: 三层或者渐变结构刨花板
7. **wood residue**: 木材加工剩余物
8. **flat hot pressing and extrusion**: 平板热压和挤出成型
9. **furnishing and construction industries**: 家装建筑行业
10. **medium-density fiberboard(MDF)***: composite panel product composed primarily of cellulosic fibers in which the primary source of physical integrity is provided through addition of a bonding system cured under heat and pressure. Additives may be introduced during the manufacturing process to improve certain properties. MDF density at the time of manufacturing is typically between 500 kg/m³ and 1 000 kg/m³, based on a reported moisture content at the time of weight and volume measurements. 中密度纤维板
11. **wet or semi-dry process**: 湿法或半干法生产工艺
12. **homogeneous structure**: 均匀的结构
13. **thermoplastic adhesives**: 热塑性树脂
14. **wood-plastic composites (WPC)***: 木塑复合材
15. **wood-thermoset composite**: 木材热固性树脂复合材料
16. **phenomenal growth**: 惊人的增长
17. **coupling agent**: 偶联剂
18. **fungal decay**: 真菌腐朽
19. **hygroscopicity*** *n* [U]: 吸湿(性)
20. **hygrothermal** *adj*: 湿热的
21. **amino resin**: 氨基树脂
22. **accelerated ageing test**: 加速老化实验
23. **hydrolytic instability**: 水解不稳定性
24. **continuing emission of formaldehyde**: 持续释放甲醛

25. **decorative laminate**: 装饰层板
26. **polyamide** *n* [C]: 聚酰胺
27. **petrochemical-based adhesive**: 石油基胶黏剂

Notes

1. 科技英语里的时态

科技英语里的时态相对比较简单, 常用的主要是一般现在时(present tense)和一般过去时(past tense)。一般现在时用于表述一般规律性的语句, 而一般过去时则表示发生在过去的某个具体的事件。如在本章中: Historically, the availability of raw materials was good and the cost was relatively low. 表示的就是过去一定历史时期的状况。而 Decorative plywood is a rigid board usually composed of an odd number of veneers. 则表示一个普适性的概念。但是有时过去时与现在时的选用也存在模糊的界线, 如某件事虽然发生在过去, 但其结果已被普遍接受, 则也可用现在时表示。如: Cellulose solubility in water is higher after treatment with compound X than with compound Y (Smith 1997; Chu et al. 1999). 就在文章中的位置而言, 过去时更多出现在论文的方法与结论部分, 而现在时主要出现在综述与讨论部分。

2. 首字母缩写与缩略语

在英文表达中, 有时为了简化拼写会将表示某个概念的若干词的第一个字母组合在一起, 如本章中的 medium-density fiberboard, 即可缩略为 MDF。缩略语(initialism)是科技英语中十分常见的一种简化表达, 在一篇文章中首次出现时必须首先给出它的完整拼写, 再在后面的括号中给出缩略语形式, 以便读者了解它的含义。是否需要采用缩略语形式主要取决于该概念的普级程度或在文章中出现的频次。如一个概念在某个领域已广为使用, 则多采用缩略语形式, 如在木材科学领域:

- EMC: equilibrium moisture content 平衡含水率
- FSP: fiber saturation point 纤维饱和点
- OSB: oriented strand board 定向刨花板

而一个概念即使不常见, 但在同一篇文章中多次出现(一般超过 5 次), 也可采用缩略语形式以简化拼写, 如:

- compressive strength along the grain (CSAG)顺纹抗压强度

在句中这类词如果前面有冠词, 其形式取决于该缩略语的第一个字母的读音(而非该字母), 如: an MOE value, 而不能写为 a MOE value。

有些缩略语可以像单词一样发音, 这种词称为首字母缩略词(acronym), 如第一章中所提到的国际林联的英文 International Union of Forest Research Organization, 可缩略为 IUFRO, 它可以像一个单词一样发音为/juːfrəʊ/, 首字母缩略词在单独使用时前面通常不加冠词, 如: members of IUFRO。

在英语文章中我们还经常遇到拉丁文的缩略语, 在科技英语里常见的有:

缩略语	拉丁文全拼	英文释义
et al.	et alii	and others, and co-workers
etc.	et caetera	and the others
e. g.	exempli gratia	for example
i. e.	id est	that is
lb	libra	scales, used to indicate the pound
vs.	versus	against
sp. (复数 spp.)	species	species(复数 species)

Exercises

1. What is the main production process of plywood?
2. What is the differences between laminated veneer lumber and plywood?
3. What are the different arrangements of particles in the extruded and flat pressed particleboard?
4. What are the differences in properties between fiberboard and particleboard?
5. What do you think are the difficulties in WPC production?
6. What are the main advantages and disadvantages of common wood adhesives?
7. What are the methods to reduce indoor formaldehyde content?

References

1. Sandberg D. Additives in Wood Products - Today and Future Development. Environmental Impacts of Traditional and Innovative Forest-based Bioproducts. Environment Footprints and Eco-design of Products and Processes. Springer Singapore, 2016.

2. Youngquist J A. Wood-based composites and panel products. Wood handbook: wood as an engineering material. Madison, WI: USDA Forest Service, Forest Products Laboratory, 1999.

3. Barbu M C, Reh R, Irle M. Wood-based composites//Research Developments in Wood Engineering and Technology. IGI Global, 2014: 1-45.

4. Low K, Burns K. Engineered wood products. Agricultural Commodities, 2013, 3(1): 130.

5. Guss L M. Engineered wood products: the future is bright. Forest Products Journal, 1995, 45(7, 8): 17.

6. Adair C. Regional Production & Market Outlook: Structural Panels & Engineered Wood Products: 2001-2006. APA——the Engineered Wood Association, 2001.

7. 周定国. 人造板工艺学. 2版. 北京: 中国林业出版社, 2011.

Wood Modification

There is an increasing demand for sustainable building materials in order to reduce CO_2 emission and energy consumption all over the world. Wood is a ubiquitous and dependable material for constructions and used in a very broad range of applications like furniture, buildings, bridges, etc. The huge diversity in timber species guarantees that there is always a species with the required properties for a specific purpose. The economic impact of timber products is therefore considerable. However, the growth of the human population is causing increasing pressure on forests with high-quality timber for construction and other purposes. Furthermore, in the last decades a significant increase of large-scale **deforestation,** especially sub-tropical forests, is observed, which contributes to the global warming, the erosion of fertile soil, and the reduction of the biodiversity. There is clearly a declining production of high quality or more specifically of durable timber from the current forests. As a consequence of the increased introduction of governmental restrictive regulations for protecting the environment, the availability of the wood material will be even more reduced. A solution to this could be the increased use of less durable timber species, which are mostly from fast-grown forests.

On the other hand, as with many naturally formed organic bio-materials, wood may be subject to **decay**, fungal infection, insects or marine borers attacked, discoloration, fire, and also **surface weathering**, which can greatly reduce the service life of wood buildings and products. Another issue is the **anisotropy of wood dimensions** caused by distribution and movement of water in wood cells. To extend the service life and ensure the quality, wood products need to be treated by appropriate preventive measures. From the perspective of the utilization of high-quality wood products, wood modification is based on the premise of retaining the inherent advantages, such as high strength-to-weight ratio, workability, sound-absorption, heat-insulation, natural texture, etc. In addition, there is a need to solve the problems of the natural inherent defects like **swelling and shrinking**, dimensional instability, anisotropy, burning, decay, discolor, etc. by a series of physical, chemical and biological methods.

To sum up, technologies for improving timber dimensional stability, fungi decay resistance, fire resistance are indispensable. A number of technologies exist, e.g. wood cell wall re-enforcement, cell lumen **impregnation**, wood surface protection, and others. Chemically modified wood is wood which has been treated by chemical processes to change the biological, chemical, mechanical or physical properties of natural wood. Reactive modification and nonreactive modification based on the most abundant functional group in wood is the **hydroxyl groups**, which follows that the reactive chemical modification of wood depends principally on reagents that readily react with these groups. The changes in wood properties which are most apparent after the chemical modification are the increase in resistance to biological deterioration and the improvement in dimensional stability.

Under the most favorable modification conditions, these improvements are achieved without sacrifice of mechanical properties or with even a slight increase in strength. One reason relates to the increased steric influence of the larger substituent compared with that of the small hydroxyl hydrogen. The other involves

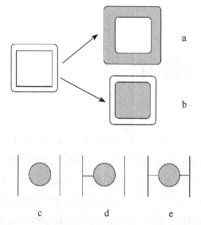

Fig.7.1 A diagram illustrating different types of wood modification at the cellular level (Source: Norimoto and Gril, 1993). Chemical modification (cell wall and surface, d, e), impregnation modification (cell wall fill, a, c; lumen fill, b), thermal modification (cell wall)

the reduced hydrogen bonding capability of the modified wood and its resultant lower **hygroscopicity**. Biological deterioration of wood, whether by fungi or by symbiotic protozoa in termites, depends on enzymatic hydrolysis of the polymeric wood carbohydrates to simple sugars. Chemical modification of wood can also provide stabilization by mechanical restraint at the molecular level by the information of cross-links between cellulose chains or microfibrils which prevent the spreading and swelling of the cellulose structure by water.

The different types of wood modification at the cellular level are concluded by Norimoto and Gril in 1993 (Fig.7.1). Chemical modification reagents mostly reacted with wood polymers in cell walls or wood surface (Fig.7.1 d, e). Most of the chemical modification methods investigated to date have involved the chemical reaction of a reagent with the cell wall polymer hydroxyl groups. This can result in the formation of a single chemical bond with one OH group or **cross-linking** between two or more OH groups (Fig.7.1e). The cell lumen modification always filled lumen or cell wall with chemicals by impregnation (Fig.7.1 a, b, c).

Cell wall modifications

Cell wall modification also can be termed as "chemical modification", which means a chemical reaction between some reactive parts of wood polymers and a simple single chemical reagent, with or without **catalyst**, to form a covalent bond between each other. Because the wood cell wall is formed by cellulose, hemicelluloses, and lignin, the most reactive site for chemically modifying is the hydroxyl group. Several basic principles during selecting a reagent and a reaction system should be considered for better improvement of the modified wood properties.

Firstly, the chemicals must have functional groups, which can react with the hydroxyl groups in wood components.

The toxicity of the reagents is another important factor needing to be considered. The chemicals must be nontoxic to humans and the natural environment when the finished products are in service and be nontoxic during the manufacturing progress.

If liquid chemicals are selected, a relatively low boiling point is advantageous for removing excess reagents after the treatment. Generally, the lowest member of a homologous series shows most reactive and will have the lowest boiling point.

Accessibility of the chemicals to the reactive sites in the wood cell wall is a critical consideration, especially for gas treatment. Multi-time accessibility of chemicals depends on the infiltration capacity of the carrier. To increase the accessibility to the reaction site, the chemical should be able to swell the wood cell wall. Otherwise, the penetrant agent or co-solvent should be added to improve the penetration of chemicals. Usually, the weak alkaline medium is the most favorable one because it can be well swelling in the cell wall matrix and lower **degradation** in wood.

Appreciate reaction time and the temperature is another important consideration. To ensure the relatively fast rate of reaction with wood cell wall polymers and reduce the degradation of wood components, it is suggested that the shorter reaction time matches up with higher temperature or the longer reaction time cooperates with lower temperature.

Wood moisture content in the reaction conditions is a key factor affecting the modification. It is costly in energy and productivity to dry wood to lower percent moisture, but the —OH group in water is more reactive than the —OH group in the wood components. The most favorable condition is a reaction which requires a trace of moisture.

Another consideration in chemical modification is to keep the reaction system as simple as possible. Multicomponent systems require complex separation after the reaction for chemical recovery. The optimum choice would be a reactive chemical that swells the wood structure that also acts as both the solvent and catalyst. If possible, avoid by-products during the reaction that must be removed. If there is not a 100% reagent skeleton add-on, then the chemical cost is higher and will require recovery of the by-product for economic and environmental reasons.

The chemical bond formed between the reagent and the wood components is of major importance. For permanence, this bond should have great stability to withstand nature's recycling system if the product is used outdoors. In order of stability, the types of covalent chemical bonds that may be formed are ethers > acetals > esters. The ether bond is the most desirable covalent carbon-oxygen bond that can be formed. These bonds are more stable than the glycosidic bonds between sugar units in the wood polysaccharides, so the polymers would degrade before the grafted ether. The hydrophobic nature of the reagent needs to be considered. The chemical added to the wood should not increase the **hydrophilic** nature of the wood components unless that is the desired property. If the hydrophilicity is increased, the susceptibility to microorganism attack and wood degradation increases. The more hydrophobic the component can be, the better the moisture exclusion properties of the substituted wood will be.

The treated wood must still possess the desirable properties of wood. Its strength should not be reduced; there should be no change in color; its electrical insulation properties are retained; the final product is not dangerous to handle; there are no lingering chemical smells; and one or more of these properties are the object of change in the product. A final consideration is, of course, the cost of chemicals and processing, especially for the commercialization of a process.

Many chemical reaction systems have been published for the chemical modification of wood. These chemicals include anhydrides such as phthalic, succinic, propionic, and butyric anhydride; acid chlorides; ketene carboxylic acids; many different types of isocyanates; formaldehyde; acetaldehyde; difunctional aldehydes; chloral; phthaldehydic acid; dimethyl sulfate; alkyl chlorides; β-propiolactone, acrylonitrile; epoxides, such as ethylene, propylene, and butylene oxide; and difunctional epoxides. Because the properties of wood resulted from the chemistry of the cell wall components, the selection of modification molecule depends on the final properties and objective of treated wood. For example, if the objective is **water repellency**, then the approach might be to reduce the hydrophilic nature of the cell wall by bonding on hydrophobic groups. If the improvement of dimensional stability is the objective, it can bulk the cell walls with bonded chemicals, or crosslink cell wall polymer components together to restrict cell wall expansion, or bond groups to reduce hydrogen bonding or increase the hydrophobicity. If fire retardancy is desired, it can bond chemical groups onto cell wall polymers that contain retardants or flame suppressants. If the resistance for ultraviolet radiation is desired, it can bond UV blockers or absorbers to lignin. The chemical modification system selected must perform the desired chemical change to achieve the desired change in performance.

Cell lumen modifications

Wood can be impregnated with other chemicals such as preservatives, polyethylene glycol or resins that improve **wood durability** and other physical properties (surface hardness, water repellency, dimensional stability, abrasion resistance, and fire resistance). In order to protect wood against biodegradation caused by bacteria, fungi, insects, termites and/or marine organism, it can be impregnated with pesticides or wood **preservatives**. In order to protect the wood from weather conditions and pest attack, it is necessary to treat it with preservatives. The preservatives used can be divided into water-based (e.g., sodium phenylphenoxide, benzalconium chloride, copper chrome arsenate); organic solvent-based (e.g., triazoles, permethrin, copper and zinc naphthenates); borates; and tar oils. Preservative treatment is a cost-effective method to protect wood against biodegradation, but the health and environmental concerns about this preservative treatment method is growing. In general, the commonly used wood preservatives (creosote and CCA) are rather poisonous for humans and animals and the treated wood possesses hazard for the environment and human health during its service and after the waste disposal (e.g. emissions of toxic chemicals). Water-soluble preservatives are beneficial to permeability but are easy to be **leached**.

Another mature method is *in situ* **polymerization** of liquid polymers in the wood lumens. Some liquid resins are vacuum-impregnated into the wood and become insoluble after curing. The polymer is located almost solely in the lumen of the wood (Table 7.1). Methyl methacrylate (MMA) and many different types of acrylics are usually used. Because wood impregnated with MMA only shows a void space at the interface between cell wall and polymer, some of the crosslinking agents are frequently used with MMA or other vinyl monomers for improving and increasing the reaction rate and physical properties of wood-polymer composites. 1,3-butylene dimethacrylate, ethylene dimethacrylate, trimethylolpropane trimethacrylate, the polar monomers 2-hydroxyethyl methacrylate, glycidyl methacrylate are the generally used crosslinking monomers. The added ratio of crosslinking monomers is often at 5% ~ 20% concentration to MMA.

The process for impregnating wood with preservatives or polymers involves drying the wood (usually at 105°C) to remove moisture, mixing or preparing reagents, and then impregnating under proper conditions. After the drying, the wood is placed in a container (large enough for the total volume of wood and reagents), and a weight is applied on the top of the wood to hold it under the solution. The vacuum and pressure are applied depending on the size of the wood to be treated. The vacuum is maintained for 5 ~ 10 minutes and then released. The pressure of 1.0 ~ 1.5 MPa is usually applied for 2 ~ 6 hours. The treated wood is removed from the solution, drained, and wiped to remove excess chemicals from the surfaces of the specimens.

Table 7.1 Properties of wood after impregnated modification

Property	Water-soluble polymers		Liquid monomers	
	PEG	Impreg	Methyl mathacrylate	Epoxy resin
Specific gravity	Slightly increased	15 to 20 pct Greater than normal wood	Increased	Increased
Permeability to water vapor	Hygroscopic	Better than normal wood	Greatly improved	Greatly improved
Liquid water repellency	Hygroscopic	Better than normal	Greatly improved	Greatly improved
Dimensional stability	80 pct	60 ~ 70 pct	10 pct	Slightly improved
Decay resistance	Better than normal	Better than normal	Somewhat increased	Somewhat increased
Fire resistance	No data	Unchanged	Unchanged	No data
Chemical resistance	No data	Better than normal	Much better than normal	Much better than normal
Compression strength	Slightly increased	Increased	Greatly increased	Greatly increased
Hardness	Unchanged	Increased	Greatly increased	Greatly increased
Abrasion resistance	Slightly reduced	Reduced	Greatly increased	Greatly increased
Machinability	unchanged	Better than normal but dulls tools	Metalworking tools preferred	Metalworking tools preferred
Glueability	Special gluer required	Unchanged	Special gluer required	Epoxy used as adhesive
Finishability	Requires polyurethane, oil, or 2 parts polymer	Unchanged	Plastic-like surface(no finish required)	Plastic-like surface(no finish required)
Color change	Little change	Reddish brown	Little change	Little change

Wood surface modification

Bulk modification can be problematic due to the difficulty of ensuring that the reagent is evenly dispersed throughout the wood material and the necessity of ensuring that all reagent and by-products are removed at the end of the treatment. However, if the treatment is confined to the surfaces of the substrate, then accessibility of reagent and subsequent clean-up of the modified material are easier to be accomplished. The properties of wood surfaces depend on the species, environmental conditions, age, and mechanical processing method. For better final use, wood needs to undergo surface improvement treatments to overcome issues resulted from surface **inactivation**, the formation of weak boundary layers, and material processing such as machining, drying, and aging. As a result, the surface modification is to protect the surface of the wood and give it a good appearance. It is also used to improve the ultraviolet stability of wood, to change the surface energy of wood (to reduce wetting by water, and/or improve compatibility with coatings or matrix materials), and to improve bonding performance. Surface modification methods are summarized in Table 7.2.

Conventional chemical modification by surface treatment is mainly used to improve the stability to weathering during wood utilized for buildings or outdoors. Generally, wood surface degrades when exposed to UV light primarily due to the instability of the lignin component. Some evidences show that the surface chemical modification with certain reagents or graft UV stabilizers leads to an improvement

Table 7.2 Summary of surface modification methods

Modification method	Main application(s)
Conventional chemical modification	Stability to weathering, compatibilization
Chemical modification with bifunctional reagent	Polymer grafting, self-bonding, stability to weathering
Surface thermoplasticization	Self-bonding
Coupling agent	Compatibilization
Chemical activation	Self-bonding
Enzymatic activation	Self-bonding
Plasma or corona discharge	Compatibilization, stability to weathering

in UV stability of wood. Acetylation is shown to be an effectively modified method against exposure to solar radiation. This is attributed to two reasons: 1) protection of the lignin phenolic hydroxyl functionalities by esterification and blocking oxidation pathways that lead to the formation of chromophoric quinonoid groups, 2) shifting the absorption maximum of lignin from 280 nm to shorter wavelengths. Butylene oxide, butyl isocyanate, linseed oil, and methacrylic anhydride are also used to coat wood surface for UV protection. Other UV stabilizers, such as dihydroxy benzophenone, hydroxyphenylbenzotriazoles, form chemical bonding with wood to provide protection in most weathering tests.

Improving wood surface hydrophobicity is another consideration for wood exposed to exterior situations. Silicone polymers reacted with wood components are formed via new Si-O-C linkages. To ensure the stable connection between wood and the siloxanes, maleic anhydride and allyl glycidyl ether are added into the reaction system. Recently, nanoparticles of metallic oxides (ZnO, TiO_2, Ai_2O_3, and Fe_2O_3) are coated on the wood surface and constructed the superhydrophobic surface and give the more functional performance of wood products.

Providing a better **bonding interface** after surface modification is an important concerned issue in wood gluability. Depending on the chemical and mechanical variations, the wood surface shows weak bonding interfaces during the composite production. Weak chemical bonding interface is the result of extractive materials migrating to the surface. Resin acids, fats/fatty acids, sterols, and waxes will render the surface hydrophobic, while sugars, phenols, tannins, and proteins will render it hydrophilic. Weak mechanical bonding interface is the result of degradation caused by light or damaged surfaces from machining processes and others. Another key factor leading to weak mechanical bonding interface is the morphology of the wood surface such as **surface roughness**, which affects wettability due to capillary force. Adhesion is affected by the presence of damaged fibers resulted from machining, as well as cracks and splits. Depending on the intended application of wood products, it is generally necessary to treat the wood to provide specific surface properties. A typical case is that softwood products used for outdoor applications need to be treated because of their degradation resulted from UV-radiation damage caused by sunlight exposure. One of the principles of wood surface activation is to generate free radicals on the wood surfaces, which provides bonding sites for resin-free board production. Fenton's reagent chemistry has been used for the direct self-bonding of wood particles. Indirect methods involve the use of surface-activated fibers in combination with other bonding agents such as furfural and lignosulphonates. Since the use of Fenton's reagent requires the presence of hydrogen peroxide, it leads to safety concerns for composite manufacturers to use such systems. Although bonding is achieved using surface activation, it is thought that, even with the 'dry' process, the moisture content of the fibers is sufficiently high to allow the diffusion of the reagents into the cell wall.

When hardwood products need to be surface-coated, it is desired to fill up the **pores** on the surface structure before applying the coating to improve the coating quality. There is increased use of wood-based board materials in a wide variety of applications. These engineered composite materials have improved properties and structural characteristics. The most common board materials are chipboards, fiberboards, and plywood.

Different materials can be used for finishing, such as fillers, stains, primers and top coatings, oils, and waxes. Fillers are used to seal **cracks** and small cavities on the wood surface. Stains are usually applied under transparent coatings with the purpose of giving color to the wood to emphasize its natural beauty and structure. Lacquers and paints are the most used coating materials and usually the outermost layer of the treated wood. Coating materials consist of binders, fillers, pigments, flatting agents, solvents, and additives. The main properties of the materials to be used for surface treatment depend greatly on the **binder**. Binder materials commonly used for the treatment of wood include amino-resins, polyurethane, acrylate, polyester, and nitrocellulose. Special properties can be achieved using a combination of these materials. Amino resins, such as urea and melamine, and alkyd resins, which are often modified with nitrocellulose, are used as binders in acid-curing treatments. These materials harden as a result of a polycondensation process that is initiated when a catalyst is added. Such a catalyst (acid hardener) should be added to the coating before application. The hardening process can be significantly accelerated by increasing the drying temperature and boosting air flow. Acid-curing coatings have good durability against chemical attack and mechanical impact.

Emulsions, colloidal dispersions, and totally water-soluble binders can be used for water-borne treatments. These products are extensively used mainly because the emission of toxic organic solvents is greatly reduced or totally avoided during the treatment. In addition, these materials have good light and fire resistance. There are a wide variety of water-borne products with different properties that can be used, depending on the application in which the wood is needed. Water-borne coatings can be alkyd-, polyurethane-, acrylate-, or polyester-based. The main drawbacks of these materials are that the storage and transportation can be carried out only at temperatures above 0°C, and the wood can swell after applying the coating.

Polyurethane materials dry as a result of a chemical reaction between the isocyanate and hydroxyl groups. Most polyurethane materials are organic solvent-based, although there are some water-based products. Compared to acid-curing materials, polyurethanes dry more slowly. Polyurethanes are known for good resistance to chemicals and mechanical impacts, as well as excellent moisture resistance. Their flexibility enables them to provide good resistance against swelling and shrinking.

Acrylate combined with a photoinitiator is generally used as a UV-curing coating. Under UV radiation, the photoinitiator starts a fast **curing** process. These materials, which can be acrylates or polyesters, need to be dried before the curing in order to remove all the solvent.

Moisture content causes wood to swell or shrink, and the results can differ depending on the wood species. The moisture content of wood should be kept stable during the treatment processes, as well as during the storage, transport, and service. Cycled or even single swelling and shrinking of the wood surface can cause more severe damage than the treated surface, resulting in cracking and splitting which appears on the surface of the coating. The coatings are usually applied to wood surfaces by pressurized impregnation, soaking, brushing, spraying, dipping/immersion or thermal diffusion (immersion in a hot bath of preservative).

The future of wood modification

There has been a dramatic increase in interest in wood modification over the past decade. The economic and environmental background is now quite different from the situation in the past when attempts were made to commercialize wood modification. The wood modification industry needs more market penetration and public profile, as well as reduced cost. Recently, the thermal modification of wood has become a new choice because of its relatively low capital expense of the equipment for processing. Compared to full chemical modification, the capital cost of plant for furfurylation is relatively low, and commercialization of this process is well developed. The introduction of a new wood modification process involves risks and it is only comparatively recently that the market has become more favourable for such ventures. Additional costs are inevitably associated with wood modification when they are compared to more conventional preservative-treated wood. At the present time, the economic cost of a product is determined by the manufacturing costs. These are determined by the cost of raw materials, capital costs of processing equipment and energy used in manufacturing and transportation, and so on. As environmental legislation has become more stringent, any part of the manufacturing process that has negative environmental impact tends to become more expensive. Forestry provides other benefits besides the production of timber. Forests, if managed correctly, also provide social benefits (tourism, leisure, and recreation); they can be used to improve the quality of water resources and to reduce flooding by reducing run-off and preventing soil erosion. By planting a suitable mix of species and utilizing appropriate management techniques, forests can be used to enhance biodiversity. Finally, of course, forests can be used to sequester atmospheric carbon. All these functions can be considered as economic benefits (so-called externalities), but it is difficult to factor these into the economics of the forestry-wood chain. If these undoubted benefits can be taken into account in the finances of timber production, this will improve the economic competitiveness of wood products. The standards that have been developed to date are concerned with determining the properties of conventional wood products. There is now an urgent need to develop appropriate standards and agree to them at an international level. At present, claims made with respect to modified wood are based upon the current set of standards, which may not always be appropriate to the performance of these new materials in real-life situations. This is an area that has to be addressed in the very near future.

Words and Phrases

1. **deforestation** *n* [U]: the conversion of forest to other land use or the permanent reduction of the tree canopy cover below the minimum 10 percent threshold. 森林采伐

2. **decay** *n* [U]: decomposition of wood substance caused by action of wood-destroying fungi, resulting in softening, loss of strength and weight, and often in change of texture and color. 木材腐朽

3. **surface weathering** *n* [U]: the mechanical or chemical disintegration and discoloration of the surface of wood

that is caused by exposure to light, the action of dust and sand carried by winds, and the alternate shrinking and swelling of the surface fibers with the continual variation in moisture content brought by changes in the weather. Weathering does not include decay. 表面老化

4. **anisotropy of wood dimensions**: the various of dimension on transverse, radial, and tangential face of wood. 木材尺寸各向异性, 表明木材是非均质材料, 是木材的重要特点之一

5. **swelling and shrinking** *n* [C]: enlargement or reduction in dimensions due to increasing or lowering the moisture content below the fiber saturation point. 湿胀与干缩

6. **impregnation*** *n* [U]: the entering of an chemical into an wood cells. 浸渍

7. **hydroxyl group*** *n* [U]: an ion with a negative charge, consisting of an oxygen atom and a hydrogen atom. 羟基官能团

8. **hygroscopicity** *n* [U]: absorbing water from the air or external environment. 吸湿性, 与木材改性密切相关

9. **cross-linking** *v*: to form chemical bonds between molecules to produce a three-dimensional network. 交联, 一种化学反应类型

10. **catalyst** *n* [U]: a substance that initiates or changes the rate of chemical reaction, but is not consumed or changed by the reaction. 催化剂

11. **degradation** *n* [U]: a reduction in quality of lumber, log, or other wood products due to processing. 降级、降等

12. **hydrophilic** *adj*: some substances can be mixed with or dissolved in water. 亲水性

13. **water repellency** *adj*: material does not absorb water. 拒水性

14. **wood durability** *n* [U]: its lasting qualities or permanence in service with particular reference to decay. 木材耐久性

15. **preservative** *n* [U]: chemical substance which, when suitably applied to wood, makes the wood resistant to attack by fungi, insects, marine borers, or weather conditions. 木材保护剂

16. **leachable** *adj*: the behavior of the modification reagents moving up from wood cells. 可析出的, 与之对应的是改性试剂在木材中的固着效率

17. **polymerization** *n* [U]: a chemical reaction in which the molecules of a monomer(s) are linked together in repeating units to form larger molecules. 聚合反应

18. **inactivation** *n* [U]: the decrease of reaction activity on wood surface. 钝化作用, 与木材表面污染、抽提物分布、加工方式等有关

19. **bonding interface** *n* [U]: a situation, form where two wood face come together by adhesive. 胶接界面

20. **surface roughness** *n* [U]: the quality of having a surface that is not even or regular or smooth. 表面粗糙度

21. **pore** *n* [C]: in wood anatomy, a term applied to the cross section of a vessel or of a vascular tracheid. 管孔, 含阔叶材的导管和导管状管胞

22. **crack** *n* [C]: the separation of the wood along the fiber direction that usually extends across the rings of annual growth, commonly resulting from stresses set up in wood during seasoning

23. **binder** *n* [U]: a component of an adhesive composition that is primarily responsible for the adhesive forces which hold two bodies together. 胶黏剂

24. **curing** *v*: to change the physical properties of an adhesive by chemical reaction, which may be condensation, polymerization, or vulcanization; usually accomplished by the action of heat and catalyst, alone or in combination, with or without pressure. 固化, 特别是人造板制造工艺中胶黏剂的固化

Notes

1. 木材改性技术

Wood modification 在本书中主要分为细胞壁、细胞腔和木材表面3个方面介绍的。但是在实际的生产与操作中,是难以将这3种改性形式完全分离开来。所以有的教材中是根据功能性改性来阐述的,如木材强化、木材尺寸稳定化、木材阻燃、木材防腐等。而且不同的功能性改良中,又以加工方式或产品类型不同而再分。

木材强化主要有:

1) impreg: 浸渍木,将水溶性低分子量树脂扩散进入木材细胞壁而使木材增容;

2) compreg: 胶压木,将树脂的前驱体扩散到板材中,高温热压过程中树脂固化而成;

3) staypak: 压缩木,在不破坏木材细胞结构的前提下,通过湿热结合压力作用使木材密实化的过程;

4) densified: 强化木,采用低熔点合金以熔融状态注入木材细胞腔中,冷却固化后与木材的复合材料;

5) wood polymers composites (WPC): 将木材单元与高分子聚合物共混,采用挤出成型的工艺生产的复合材料。

2. 英语中的汉语表达

汉语的很多词汇无法通过意译的方法转化为英文,在科技英语中最常见的是人名、地名或物名,此时在英语中通常按照汉语拼音(Hanyu Pinyin romanization system)拼写规则进行书写。如:

- Zhang Shan 张三
- hongmu 红木

在表示人名时姓和名应按照汉语中原有的顺序,即姓在前,名在后,而非遵从西方的姓名表达方式,除非该姓名已经西化,如: Jack Ma。

除拼音系统外,还有两个目前已废止的字母化汉语表达方式,分别是魏妥玛拼音(Wade-Giles system)和邮政拼音(Postal Atlas of China),但是因为长期使用习惯等原因,它们在某些特定的场合仍获得保留,如:

- Tsinghua University 清华大学
- Soochow University 苏州大学
- Yangtze River 长江

Exercises

1. Understand the relationship between wood modification and wood cell wall structure.

2. From the anatomical structure, what are the main channels for transporting of modification reagents?

3. From the chemical structure standpoint, please explain why the lignocellulosic material presents hygroscopic characteristics.

References

1. Zeng W, Tomppo E, Healey S P, et al. The national forest inventory in China: history-results-international context. Forest Ecosystems, 2015, 2:23.

2. Chinese Forestry Society, National Poplar Commission. Forest Resource, Timber Production and Poplar Culture in China. In: Proceedings of International Conference on the future of poplar, Rome, Italy, November 13-15, 2003.

3. Shmulsky R, Jones P D. Forest products and wood science an introduction (Sixth Ed). John Wiley & Sons, Inc. Published, 2011.

4. 李坚. 木材科学. 3版. 北京: 科学出版社, 2014.

Wood for Energy Production

Wood as an energy resource

Wood has long served as a principal energy source for human beings. In 2015, wood fuel production accounted for 51% of all roundwood produced. Globally, biomass power capacity was approximately 14 GW in 2000. Energy is still the No. 1 product from forests.

There are 5 sources of woody **biomass** for energy generation. The first source is roundwood from growing stock. As mentioned above, more than half of the global roundwood is used as fuels. In some developing areas, the figure can be as high as 90 percent.

Many wood fuels come from residues of manufacturing processes. The portion of the woody biomass not used in the manufacturing processes is termed residue. In wood processing factories, residues include **planer shavings**, barks, slabs, trim, sawdust, and sanding dust. The flammability and universal availability of wood residues make them the natural choice for the forest industry to provide energy for the manufacturing processes. To produce 1 m^3 **dressed pine lumber**, as much as 55 kg of planer shavings can be formed as by-products. The heat in these shavings is around 750 MJ. Given that the heat to dry 1 m^3 green lumber is approximately 1.4 GJ, and more than half of the fuel for the dry kilns can come directly from the production process. For some factories, the energy produced by wood residues is even more than its consumption, making them 100% energy self-sufficient.

Logging residues may be another source of wood energy, which includes tops, limbs, thinning, etc. But their utilization is limited by the costs for collection and transportation. Concerns have also been raised about the ecological effects of the removal of logging residues. Many believe that they are of great importance for forest wildlife, erosion control, and nutrient cycling.

Urban and industrial wastes, including wood **pallets**, **railroad ties**, demolition wood, and urban tree removals, also present a significant source of energy. These materials have high heat value because they are most often dried. Those that are not preservative treated are clean for burning. The main challenge for industrial scale utilization is to find a cost-effective way to collect and sort them.

On weight or volume base, wood is not the optimal choice of energy source (Table 8.1). But wood has the advantage of low cost, which is one of the vital factors the industry concerns. Many wood fuels are considered solid wastes if not utilized on site. Besides, wood is a renewable energy source and is environmentally friendly by nature. The supply of wood is sustainable if the forests are managed scientifically. The burning of wood is a carbon neutral process, i. e., the net increase of **greenhouse gas (GHG)** is zero given that trees absorb **carbon dioxide** for the **photosynthesis**. Wood has low sulfur and nitrogen contents. This makes the burning cleaner than fossil fuels such as coal, which generate a significant amount of sulfur dioxide and nitrous oxides during burning. The major emission from wood burning is **particulate matters** (PM), which are mainly the unburned solid contents of wood, and can be effectively controlled by increasing the combustion efficiency and installation of proper dust collectors such as bag filters and wet scrubbers.

Table 8.1 Energy content comparison between wood and some fossil fuels

Fuel type	Energy content(GJ/t)
Wood (planer shavings, 15%MC)	17
Wood (sawdust, 90%MC)	10
Coal	23
Natural gas	42

(source: Bowyer, 2003)

The amount of heat per **oven-dry** kilogram varies little among species. It is the moisture content (MC) that greatly influences the amount of heat from wood. It is easy to understand that higher MC means lower heat value (Table 8.2) and a lower market price. In this respect, mill residues are good energy sources because they are dried in most cases.

Table 8.2 Influence of moisture content on hardwood gross and usable heat

Moisture content (%)	Gross heat (GJ/t)	Combustion efficiency (%)	Usable heat (GJ/t)
0	19.2	80	15.4
15	16.7	78	13.0
30	14.8	76	11.3
60	12.0	72	8.6
100	9.6	67	6.4

(source: Bowyer, 2003)

Although some wood residues are burned as received, it is a common practice to dry wood before using it as fuel. The drying is in most cases accomplished by heating the wood to vaporize the moisture, and various dryers are available to be selected, such as tube, drum, cascade or screen dryer. The heat for these dryers is typically from fines of the processed fuels, flue gas or other low-cost sources.

Dried wood fuel presents not only high heat value but also increased **boiler** capacity. Lower moisture contents mean higher boiler efficiency, less water vapor generation, and lower flue gas volume. This allows an increase in the amount of wood that can be burned and heat that can be produced. The steam generated by burning the fuel with MC of 40% is around 10% more than the fuel with MC of 50%. Likewise, a smaller boiler can be installed if drier fuels are adopted. Lower flue gas volume means less operating costs of boilers. Studies showed that an increase of fuel MC from 52% to 63% doubled the particulate emission rate. It is therefore desirable to burn drier fuels because they bring economic, environmental and quality benefits.

In the wood processing industry, the energy generated by wood combustion is mainly in the form of heat. A great part of it is used for the drying process. For a lumber manufacturing plant, the share can be as high as 70%. It is also possible to generate electricity from woody biomass, but only large mills can make this process economic practical. A process call **cogeneration** can produce electricity and steam simultaneously. In this case, a high-temperature and high-pressure boiler is installed, plus a **turbine** that generates electricity from the high-pressure steam and supplies heat from the low-pressure steam (Fig.8.1).

Cogeneration is appealing to some firms as most of the steam can be utilized. An efficient steam turbine power plant requires about 10 MJ to produce 1 kW · h (3.6 MJ) of electricity. For each kWh generated, as much as 5.3 MJ in the steam can be extracted to supply heat for the manufacturing processes. Therefore, the energy conversion efficiency of the power plant is only around 35%, while the cogeneration system can raise the number to about 90%.

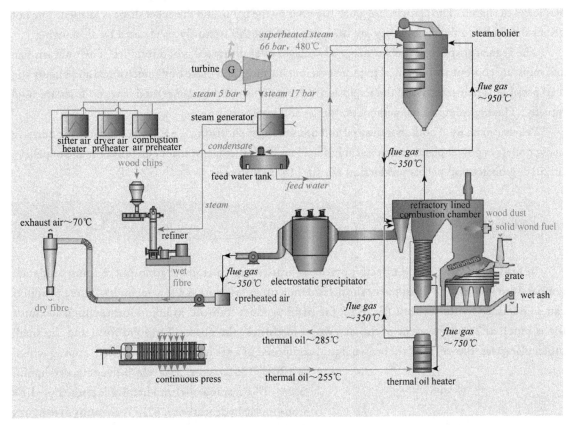

Fig.8.1 A cogeneration system for an MDF plant (source: Dieffenbacher)

A traditional utilization as it is, the way we produce energy from wood is totally different from the conventional method. Besides combustion in the form of fines, **pellet** and **briquette**, **pyrolysis**, **gasification** and liquid fuels produced in **biorefinery** are choices available.

Pellet

Pellets are cylinder-shaped biofuel particles produced by compressing biomass particles. The standard dimension of pellet fuel is 6 ~ 8 mm in diameter and less than 38 mm in length. Larger pellets are usually referred to as briquettes. Wood pellets are the most common type of pellet fuel and are generally made from sawdust and other wood processing wastes. The process of manufacturing pellets involves feedstock grinding, moisture control, extrusion, cooling, and packaging. In standard-sized pellet mills, the raw material is firstly chipped and ground to particles of less than 3 mm in size. Moisture content is a key performance indicator of pellets. It influences not only the processing properties but also the combustion performance of the final products. For wood, the moisture content is kept at around 10%. The drying is performed by either the oven-drying method or blowing hot air over or through the particles. A too dry state of the particles should also be avoided. If this occurs, particles are re-wetted by injecting steam or water into the feedstock. The biomass is then fed to a press, where it is squeezed through a round opening called a "die" with the required size of the final product. The press rises wood temperature and subsequently plasticizes the lignin. The particles are glued together by the lignin and the pellet shape is formed. The hot pellets should be cooled before they are ready for use, and this is usually accelerated by air blowing.

The regular shape gives pellets many advantages over untreated wood residues. The uniform size and moisture content are beneficial to more efficient burning, and make both mechanical and pneumatic feed possible. Their sizes avoid the explosion danger that should be concerned if wood fines are used directly. The high-density conveniences storage and transportation.

Pellets can be used both in industrial and household applications, i. e. they are suitable for burning in big boilers in power plants and stoves at home. This is another factor that helps boost the use of pellets. In 2010, global wood pellets production reached 14 million tons.

Pyrolysis of wood

Instead of burning wood directly, pyrolysis heats wood in an inert atmosphere using moderate temperature (400 ~ 500 ℃) and very short residence times (as short as a few seconds). Three products can be obtained by this process (Fig.8.2): the main product 'bio-oil' in liquid form, which accounts for as much as 75% by weight of the dry wood feedstock, the charcoal in solid form and the small molecular gases. Bio-oil is a dark brown liquid composed of a great variety of chemicals, among which are 20% ~ 25% water, 25% ~ 30% decomposed lignin, 5% ~ 12% organic acids from hemicelluloses, 5% ~ 10% non-poplar hydrocarbons, 5% ~ 10% **anhydrosugars** from **polysaccharides** and 10% ~ 25% other compounds. The heat value of bio-oil is around 17 MJ/kg, which makes it a potential substitute for fuel oil used in boilers, gas turbines and diesel engines. The gas product of the pyrolysis also has heating value and is often burned directly in the pyrolysis reactor to provide the reaction heat. The **charcoal** is a good raw material to prepare active carbon. Successful pyrolysis requires the feedstock to be in a dry condition. The moisture content is supposed to be less than 10% to facilitate the grinding of wood.

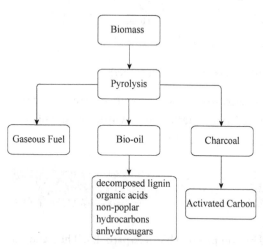

Fig.8.2 Products of biomass pyrolysis

Wood gasification

Gasification is a process that converts biomass into so-called **syngas**, a mixture of hydrogen, **carbon**

monoxide, carbon dioxide, **methane** and **hydrocarbon,** and **tar** components. This is achieved by heating the feedstock at temperatures over 700 ℃ in an environment with limited oxygen to guarantee a gaseous product with heating value as high as 20 MJ/Nm^3.

The gasification process can be divided into 4 phases, i. e. drying, pyrolysis, gasification, and combustion. At temperatures under 100 ℃, moisture evaporates from the wood. The resulting steam is mixed into the gas and will be involved in the subsequent reactions. When the temperature climbs to 200 ~ 300 ℃, hydrogen, carbon monoxide, carbon dioxide, methane, and hydrocarbon are produced. What left in the solid form is char, which can react with oxygen and water to generate more carbon oxides and hydrogen at higher temperatures. The relative amounts of the reaction products reach an equilibrium state by a water-gas shift reaction, where the concentrations of carbon monoxide, steam, carbon dioxide, and hydrogen are balanced. The reaction can be summarized by Formula (8-1):

$$[CH_{1.44}O_{0.66}]+H_2O+O_2=H_2+CO+CO_2+H_2O+CH_4(+tar/hydrocarbons) \tag{8-1}$$

Liquid fuels

Wood is a **heterogeneous** material that can be transformed into liquid fuels, such as **methanol**, **ethanol** or Fischer-Tropsch liquids (a mixture of hydrocarbons). Alcohol can substitute for gasoline while one of the main Fischer-Tropsch liquid products is diesel.

Two well-developed methods exist for manufacturing liquid fuels from wood. The syngas from the gasification process can be used as the raw material for producing liquid fuels. In this process, the syngas is passed over metal catalysts at temperatures of 150 ~ 300 ℃ and pressures of up to tens of atmospheres. The products obtained are methanol and Fischer-Tropsch liquids. The other method is **fermentation** of sugars, which are obtained from cellulose and hemicelluloses in the wood by **hydrolysis**. The final product is ethanol. In the United States, ethanol produced this way mainly comes from corn. But it is not economically sustainable without the subsidies from the government. Brazil, meanwhile, adopts sugar cane to produce ethanol by fermentation, which has proven to be more cost-effective compared to corn-based ethanol. One advantage of ethanol is that it can be blended with gasoline and used in automobiles directly. It is also less polluting than most fossil fuels. The main obstacle to its wide application is still economic feasibility. In the United States, for example, the emerging **shale gas** production technology has sharply dropped the fossil-based energy price and made bio-based alternatives much less concerned.

Ethanol can also be produced by fermenting the syngas. The advantage of the latter is that lignin can also be used as raw material while the former only utilizes the carbohydrates components in the wood.

Very limited commercial adoption of biofuels from wood is available worldwide. The prospect of it depends on its production cost compared to fossil fuels, environmental concerns about air quality and global warming, and the sustainability of its supply as a renewable material.

Words and Phrases

1. **biomass*** *n* [U]: natural materials from living or recently dead plants, trees and animals, used as fuel and in industrial production, especially in the generation of electricity. 燃料用生物质

2. **planer** *n* [C]: 刨床

3. **shaving*** *n* [C]: the thin slices of wood removed in dressing. (刨切)刨花

4. **dressed lumber***: lumber that is surfaced by a planning machine on one side (S1S), two sides (S2S), one edge (S1E), two edges (S2E), or any combination of sides and edges (S1S1E, S2S1E, S1S2E, or S4S). Dressed lumber may also be referred to as planed or surfaced. 刨光板

5. **pallet** *n* [C]: 托盘

6. **railroad tie**: 枕木

7. **greenhouse gas(GHG)**: 温室气体

8. **carbon dioxide**: 二氧化碳

9. **photo synthesis**: 光和作用

10. **particulate matter(PM)**: 颗粒物

11. **overdry*** *adj*: dried in an oven to remove all moisture. The temperature employed usually is 101 to 105°C or 214 to 221°F in accordance with ASTM Methods D 2016 *Test for Moisture Content of Wood*. 绝干的

12. **boiler** *n* [C]: 锅炉

13. **cogeneration*** *n* [U]: 热电联产

14. **turbine** *n* [C]: 涡轮

15. **pellet*** *n* [C]: 颗粒燃料

16. **briquette** *n* [C]: 块状燃料

17. **pyrolysis** *n* [U]: 热解

18. **gasification** *n* [U]: 气化

19. **biorefinery** *n* [C]: a facility that processes biological material (such as crop waste) to produce fuel (such as ethanol and biodiesel), electricity, and commercially useful chemicals (such as succinic acid). 生物质精炼厂

20. **anhydrosugar** *n* [C]: 脱水糖

21. **polysaccharide** *n* [C]: 多糖

22. **charcoal** *n* [U]: 木炭

23. **syngas** *n* [C]: synthesis gas 的简写,合成气

24. **carbon monoxide**: 一氧化碳

25. **methane** *n* [U]: 甲烷

26. **hydrocarbon** *n* [C]: 碳水化合物

27. **tar** *n* [U]: 焦油

28. **heterogeneous** *adj*: 异质的

29. **methanol** *n* [U]: 甲醇

30. **ethanol** *n* [U]: 乙醇

31. **fermentation** *n* [U]: 发酵

32. **hydrolysis** *n* [U]: 水解

33. **shale gas**: 页岩气

Notes

可数名词与不可数名词

英语里的名词可分为可数名词和不可数名词两类。可数名词表示可以计数的人和物,在表示单个个体时前面可以加不定冠词 a/an,当表示多个数量的集合前面可加数字,并具有复数形式,如:

- a panel
- an MDF sample
- two boards
- three dry kilns

不可数名词表示无法计数的抽象概念或难以单独分离，通常被视为一个整体的物质等，它们多数以单数形式进行使用，前面也不加表示数量的数字，如：

- pyrolysis
- water
- climate

多数情况下可数名词和不可数名词是容易识别的，但是在英语中也有一些不可数名词可能会被当作可数词而出现使用错误，如 furniture 表示家具，我们不能说 a furniture 或 two furnitures，只能说 a piece/an article of furniture，但是我们可以说 a table, a chair。同样地，我们不能说 a lumber 或 a timber，但我们可以说 a board, a panel。因此区分一个词是可数还是不可数最好在字典中进行核实。

有一些词既可以做可数名词也可以做不可数名词，这取决于它们所表示的含义，对于表示材料的单词而言，当其表示材料的概念时一般为不可数名词，而当其表示由该材料构成或制成的物质时可用作可数名词，如 glass 表示玻璃时为不可数名词，而表示玻璃杯时则为可数名词；paper 表示纸张时为不可数名词，而表示论文时则为可数名词；wood 在表示木材时通常为不可数名词，但表示不种类时则为可数名词，等等。试比较：

- All the furniture was made of wood/a variety of woods.
- Teak is a hard (kind of) wood and pine is a soft (kind of) wood.

Exercises

1. What are the main benefits of using wood to generate energy?

2. Compare the biofuel production methods introduced in this chapter. Which one do you think has the greatest potential for future application? Why?

References

1. Food and Agriculture Organization of the United Nations. 2015 Global Forest Products Facts and Figures. FAO, 2015.

2. Bowyer J L, Shmulsky R, Haygreen J G. Forest Products and Wood Science, An Introduction. 4th edition. Ames: Blackwell Publishing Company, 2003.

3. Walker J. Primary Wood Processing Principles and Practice. 2nd edition. Dordrecht: Springer, 2006.